ENGINEERING AND GENERAL GEOLOGY

ENGINEERING AND GENERAL GEOLOGY

Anu Mehta

ANMOL PUBLICATIONS PVT. LTD.
NEW DELHI-110 002 (INDIA)

ANMOL PUBLICATIONS PVT. LTD.
Regd. Office: 4360/4, Ansari Road, Daryaganj,
New Delhi-110002 (India)
Tel.: 23278000, 23261597, 23286875, 23255577
Fax: 91-11-23280289
Email: anmolpub@gmail.com
Visit us at: www.anmolpublications.com

Branch Office: No. 1015, Ist Main Road, BSK IIIrd Stage
IIIrd Phase, IIIrd Block, Bangalore-560 085 (India)
Tel.: 080-41723429 • Fax: 080-26723604
Email: anmolpublicationsbangalore@gmail.com

Engineering and General Geology

First Edition, 2010

ISBN 978-81-261-4588-1

PRINTED IN INDIA

Printed at Mehra Offset Press, Delhi.

Contents

Preface

Field work, supplemented by laboratory studies, is a cornerstone for the geological sciences. Even with the advent of remote sensing techniques that enable geoscientists to interpret many features using high-altitude or space photography, visits to the field are still required for the purposes of ground truthing. It is essential that geologists become familiar with field methods in their special area of interest. And it is also important for those working in laboratories to have at least rudimentary knowledge of field conditions and sampling procedures because these often affect, either directly or indirectly, interpretation of lab data.

The textbook Engineering and General Geology provides an introduction to general field work through selected topics that illustrate specific techniques and methodologies. One hundred and twenty-three main entries prepared by leading authorities from around the world deal with aspects of exploration surveys, geotechnical engineering, environmental management. field techniques, mapping, prospecting, and mining.

Author

Chapter 1

Minerals

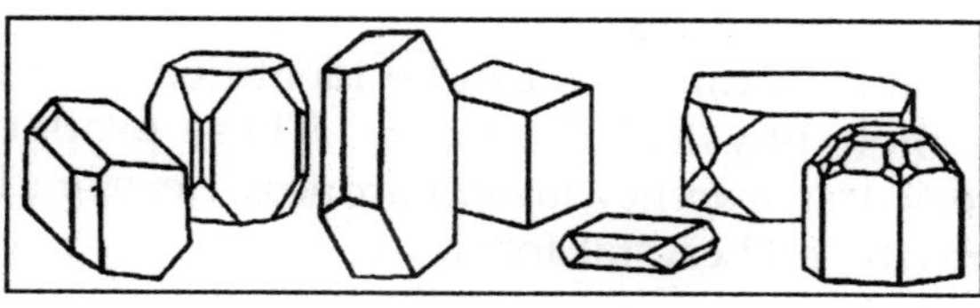

DEFINITION

The following definitions on what constitutes a mineral were taken from several different sources and are arranged by year:

- "A mineral is an element or chemical compound that is normally crystalline and that has been formed as a result of geological processes.
- "Minerals are naturally-occurring inorganic substances with a definite and predictable chemical composition and physical properties.".
- "A mineral is a naturally occurring homogeneous solid, inorganically formed, with a definite chemical composition and an ordered atomic arrangement".
- "These... minerals ...can be distinguished from one another by individual characteristics that arise directly from the kinds of atoms they contain and the arrangements these atoms make inside them".
- "A mineral is a body produced by the processes of inorganic nature, having usually a definite chemical composition and, if formed under favourable conditions, a certain characteristic atomic structure which is expressed in its crystalline form and other physical properties".

- "Every distinct chemical compound occurring in inorganic nature, having a definite molecular structure or system of crystallization and well-defined physical properties, constitutes a mineral species".

IDENTIFYING MINERALS

The first thing to have when identifying minerals is a good field guide. With your field guide in hand, you will be able to compare the physical properties of a mineral to descriptions and pictures.

Most common minerals can be identified by inspecting or testing their physical properties. These properties are colour, streak, transparency,luster,hardness cleavage, fracture, specific gravity, and crystal form.

Colour

Usually, we notice the colour of a mineral first. Some minerals are easily identified by colour because they are never any other colour. For example, malachite is always green. Keep in mind, however, that colour by itself isn't enough to identify a mineral. Chemical impurities can change the colour of a mineral without changing its basic make-up.

For example, quartz in its purest form is colourless and clear as glass. Quartz with traces of iron becomes violet (amethyst). With traces of manganese, it turns pink (rose quartz). If quartz is exposed to radiation, it turns brown (smoky quartz).

Streak

When a mineral is rubbed firmly across an unglazed tile

of white porcelain (a streak plate), it leaves a line of powder. This is called the streak. The colour of the streak is always the same, whether or not the mineral has impurities. For example, quartz leaves a white streak, whether it's violet (amethyst), pink (rose quartz), or brown (smoky quartz).

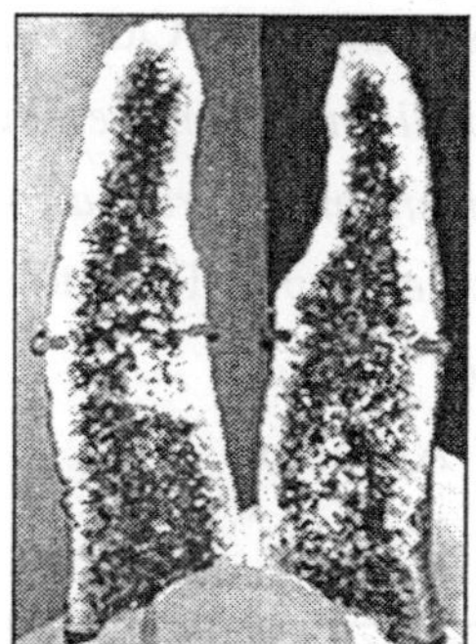

Fig. The Geode Pictured here is Lined with Amethyst Crystals.

Transparency

Transparency describes how well light passes through a mineral sample. There are three degrees of transparency: transparent, translucent, and opaque. You can see objects through a transparent mineral. You can see light, but no objects through a translucent mineral. You can't see anything through an opaque mineral.

Fig. The sample of Beryl shown here is Nearly Transparent.

Luster

Luster is the way the surface of a mineral reflects light.

Luster should be observed on a cut or freshly broken, untarnished surface. There are two general types of luster — metallic and non-metallic.

The terms used to describe luster are:

- *Metallic example*: Gold
- *Vitreous (glassy) example*: Quartz, tourmaline
- *Adamantine (brilliant) example*: Diamond
- *Resinous (like resin or sap from a tree) example*: Sphalerite
- *Greasy or waxy example*: Turquoise
- *Pearly example*: Talc
- *Silky example*: Asbestos
- *Dull or earthy example*:

Fig. Bauxite Tourmanline has a Vitreous (glassy) Luster.

Hardness

The hardness scale was established by the German mineralogist, Friedrich Mohs. The Mohs' hardness scale places ten common or well-known minerals on a scale from one to ten. One is the softest mineral and ten is the hardest. These are the minerals used in the Mohs' hardness scale:

Table. Mohs' Hardness Scale

1	2	3	4	5	6	7	8	9	10
Talc	Gypsum	Calcite	Fluorite	Apatite	Feldspar	Quartz	Topaz	Corundum	Diamond

To use the hardness scale, try to scratch the surface of an unknown sample with a mineral or substance from the hardness scale (these are known samples). If the unknown sample cannot be scratched by feldspar but it can be scratched by quartz, then it's hardness is between 6 and 7. An example of a mineral that has a hardness between 6 and 7 is pyrite (6 to 6.5).

If you don't have minerals from the hardness scale on hand, here are some common objects and their hardness values:

Table. Common Objects and Their Hardness Values

2.5	3.5	5.5	6.5	8.5
Fingernail	Penny	Glass	Steel knife	Emery cloth

If an unknown sample can not be scratched by your fingernail (2.5) but it can be scratched by a penny (3.5), then it's hardness is between 2.5 and 3.5. An example of a mineral that has a hardness between 2.5 and 3.5 is calcite (3).

Cleavage

When a mineral sample is broken with a hammer, it breaks along planes of weakness that are part of its crystalline structure. These breaks are cleavages. Some minerals break only in one direction. Others break in two or more directions.

Some common forms of cleavage are cubic, rhombohedral, and basal. Cubic cleavages form cubes (example, halite). Rhombohedral cleavages form six-sided prisms (example, calcite). Basal cleavages occur along a single plane parallel to the base of the mineral (example, topaz).

If a mineral breaks easily and cleanly in one or more directions, its cleavage is considered perfect. For example, calcite cleaves perfectly along three planes. As the quality of the break decreases, cleavage may be described as good, distinct, and poor or none. Some minerals cleave perfectly in one direction and poorly in others. For example, gypsum cleaves perfectly on one plane and poorly along two others.

Fracture

Not all minerals cleave easily. Some fracture instead.

Unlike cleavages, which are usually clean, flat breaks, fractures can be smoothly curved, irregular, jagged or splintery. The most common types of fracture are conchoidal (quartz), fibrous or splintery, hackly (copper), uneven or irregular.

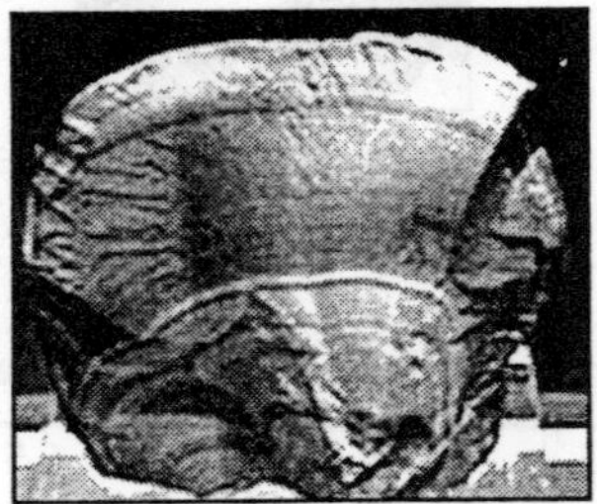

The sample of malachite shown here has a conchoidal fracture; it's smooth and curved.

Specific Gravity

Specific gravity is the density of a mineral. Special equipment is usually needed to find out a mineral's exact specific gravity. With a little practice, you can guess a mineral's specific gravity by hand. Some mineral samples will feel heavier than others, even if all your samples are the same size. The heavier ones have a greater specific gravity.

Table. Here are some Examples of Common Minerals and their Specific Gravity Ranges:

Minerals	Density	Specific gravity
Sulfur, graphite	light	1-2
Gypsum, quartz	medium	2-3
Fluorite, beryl	medium heavy	3-4
Corundum, most		
Metal oxides	heavy	4-6
Native gold, platinum	heaviest	19

Crystal Form

Minerals grow in specific shapes, and usually crystallize into one of six crystal systems. The axes of the crystal, the angles at which the axes intersect, and the degree of symmetry define each system.

- *Isometric*: Also called the cubic crystal system. Crystals are usually shaped like blocks, with similar and symmetrical faces. The crystal has three axes of symmetry, all at right angles to each other, and all of the same length.

Example: Pyrite

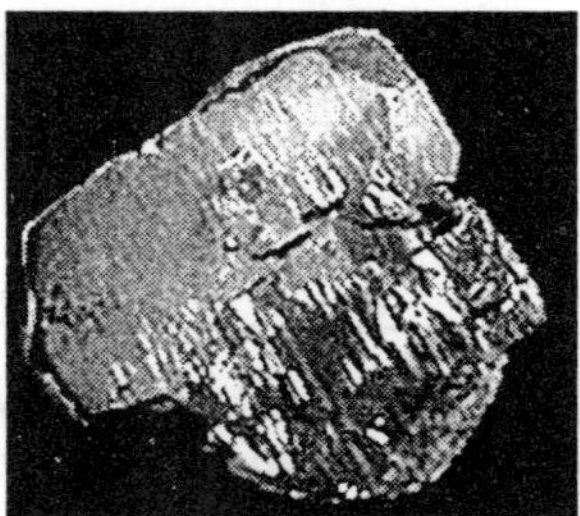

Fig. Pyrite

Description and Occurrence

Pyrite crystal usually forms as a cube with fine ridges (striations) on the crystal's faces. Less commonly, it forms as octahedrons (eight-sided shapes), nodules, or massive forms. It can also occur as coarse granules.

Pyrite is called fool's gold because its brassy yellow colour is very similar to gold. Although it looks like gold, its other physical properties are very different. Pyrite is harder, less dense, and more brittle. It leaves a greenish-black streak while gold leaves a golden-yellow one. However, pyrite is often associated with the presence of gold and copper, and locating fool's gold may mean the real thing isn't far off.

Pyrite is the most common of the sulfide minerals and can be found worldwide. It's the most important source of sulfur after native sulfur.

The mineral has a brassy yellow colour. It tends to form in cubes. The crystal faces are striated (lined with fine ridges). Striking it with steel produces a spark.

Physical Properties

- *Tetragonal*: Typically, the crystals are shaped like

four-sided prisms and pyramids. Each crystal has three axes, all perpendicular to one another. Two axes are the same length and lie on a horizontal plane. The third axis is not the same length and is at a right angle to the other two.

Example: zircon.

Color	Streak	Transparency	Luster	Hardness	Cleavage	Fracture	Specific gravity	Crystal for m
pale yellow to brass yellow	greenish black	opaque	metallic	6 to 6.5	imperfect	conchoidal, brittle	4.9 to 5.2	isometric

- *Hexagonal*: These crystals are usually shaped like six-sided prisms or pyramids. Each crystal has four axes of symmetry. Three lie in the same plane, are the same length, and intersect at 120° angles. The fourth axis is not the same length, and is perpendicular to other three.

Example: Beryl

Description and Occurrence

Beryl is a common crystal and forms as a six-sided prism. Individual crystals can become very large, as long as 9 metres and weighing as much as 25 tons. The crystal face shows very fine striations (ridges) along its length. It usually occurs in granite rocks and pegmatites.

Beryl crystals come in a wide range of colours. The most highly prized form of beryl is the emerald (medium to dark toned green or bluish-green). Other beryl gems include aquamarine (light blue and light bluish-green), golden beryl (golden yellow), heliodor (yellow and brown), morganite (pink to light purplish-red), and goshenite (colourless).

Most beryl is not gemstone quality, and is used as the main industrial source for beryllium. Beryllium is used in the nuclear industry and in alloys for aircraft.

Beryl can be found in Columbia, Austria, Brazil, India, Australia, and the United States. In San Diego County, gemstone quality aquamarine and morganite have been found in pegmatites in Mesa Grande, Pala, and Ramona. Best field marks are its six-sided shape and hardness.

Physical Properties

Color	Streak	Transparency	Luster	Hardness	Cleavage	Fracture	Specific gravity	Crystal form
bright green, blue, greenish-blue, yellow, red, pink, white, colorless	white	transparent to opaque	vitreous	7.5 to 8	indistinct, one direction	conchoidal, uneven	2.6 to 2.9	hexagonal

Orthorhombic: These crystals are short and stubby. Each crystal has three unequal axes, all at right angles to one another.

Example: Topaz

From the Greek, *topazos,* for any stone that has lost its name.

Fig. Topaz

Description and Occurrence

Topaz crystals are usually shaped like prisms, with fine vertical ridges (striations) on the crystal faces. The crystals come in a variety of colours: clear, yellow, blue, green, violet, or reddish brown. The most highly prized colour is a deep, golden-yellow variety, though blue and green stones are popular also.

Until the mid-nineteenth century, topaz was considered a rare and valuable gemstone. When rich fields of topaz were discovered in Brazil, the value of the mineral collapsed. Today, Brazil is still considered the world's best source of topaz. Topaz also occurs in Sri Lanka, Mexico, Japan, and Russia.

Fine samples have been found in Little Three Pegmatite, near Ramona in San Diego County

Field Notes: Topaz is very hard and translucent (light passes through it). It cleaves easily and perfectly on a plane parallel to its base.

Physical Properties

Color	Streak	Transparency	Luster	Hardness	Cleavage	Fracture	Specific gravity	Crystal form
colorless, yellow, brown, blue, green	colorless	transparent to translucent	vitreous	8	perfect, one basal direction	conchoidal, uneven	3.5 to 3.6	orthorhombic

- *Monoclinic*: Crystals are short and stubby with tilted faces at each end. Each crystal has three unequal axes. Two axes lie in the same plane at right angles to each other. The third axis is inclined.

Example: Gypsum

From the Greek, *gypsos*, meaning "plaster."

Fig. Gypsum

Description and Occurrence

Gypsum is an evaporite, which means its crystals form during the evaporation of water. The crystals are shaped like prisms or flat plates, and can grow up to 1 metre. Gypsum can appear as transparent crystals (selenite); fibrous, elongated crystals (satin spar); granular and compact masses (alabaster); and in rosette-shaped aggregates called desert roses.

Most gypsum is used in the building and agricultural industries. As a building material, it's used in plaster, plaster of Paris, wallboard, cement, and ceramic tiles. In agriculture, it's used as an amendment to neutralize alkaline soil. Some gypsum — dense and fine-grained — is called alabaster and can be carved.

Common around the world, gypsum is found primarily in sedimentary rock. In North America, crystals can be found in New York, Utah, and Oklahoma. In this region, gypsum is mined and processed in a major production plant, located in the aptly named Plaster City between Ocotillo and El Centro. The whole area is white with dust.

Gypsum clusters, called desert roses, form in the desert from the evaporation of groundwater. Desert roses can be found near Ocotillo, in the Imperial Valley.

During the Museum's exhibit of Shona sculpture, the artist tried carving with alabaster. The piece turned out magnificently; however, the artist did not like the softness of the material. Gypsum is very soft, and the crystals are usually flat. Desert roses often incorporate grains of sand.

Physical Properties

Color	Streak	Transparency	Luster	Hardness	Cleavage	Fracture	Specific gravity	Crystal form
colorless, white, gray, yellow, red, brown	white	transparent to opaque	vitreous, pearly on cleavage	1.5 to 2	perfect in one direction	conchoidal, splintery	2.3 to 2.4	monoclinic

- *Triclinic*: Crystals are usually flat with sharp edges, but exhibit no right angles. Each crystal has three unequal axes. None are perpendicular to one another.

Example: Feldspar.

CRYSTALLOGRAPHY AND MINERAL CRYSTAL SYSTEMS

Crystallography is a fascinating division of the entire study of mineralogy. Even the non-collector may have an appreciation for large well-developed beautifully symmetrical individual crystals, like those of pyrite from Spain, and groups of crystals, such as quartz from Arkansas or tourmaline from California, because they are esthetically pleasing. To think that such crystals come from the ground "as is" is surprising to many.

The lay person simply hasn't had the opportunity to learn about crystals and why they are the way they are; however, neither have many rock hounds and hobbyists. We hope to bring you to a greater appreciation of natural mineral crystals and their forms by giving you some background and understanding into the world of crystallography.

CRYSTALLOGRAPHY is simply a fancy word meaning "the study of crystals". At one time the word crystal referred only to quartz crystal, but has taken on a broader definition which includes all minerals with well expressed crystal shapes.

Crystallography may be studied on many levels, but no matter how elementary or in-depth a discussion of the topic we have, we confront some geometry. Oh no, a nasty 8-letter word! Solid geometry, no less.

But stop and think about it, you use geometry every day, whether you hang sheetrock, pour concrete, deliver the mail, or work on a computer. You just don't think of it as geometry. Geometry simply deals with spatial relationships. Those relationships you are familiar with are not intimidating. The key word here is "familiar". We want this series of articles to help you become more familiar, and, therefore more comfortable, with the geometry involved with the study of crystal forms.

Crystallography is easily divided into 3 sections — geometrical, physical, and chemical. The latter two involve the relationships of the crystal form (geometrical) upon the physical and chemical properties of any given mineral. We will cover the most significant geometric aspects of crystallography and leave the other topics for later.

We do not intend for this series to be a replacement for a mineralogy textbook, but instead an introduction to the study of crystallography.

In any type of study, there exists special words used to summarize entire concepts. This is the special language of the "expert", whether you speak of electrical engineering, computer science, accounting, or crystallography. There's no real way to get around learning some of these basic definitions and "laws" so we might as well jump right into them.

First, let's define what we're dealing with. A CRYSTAL is a regular polyhedral form, bounded by smooth faces, which is assumed by a chemical compound, due to the action of its interatomic forces, when passing, under suitable conditions, from the state of a liquid or gas to that of a solid. WOW, what a mouthful! Let's dissect that statement.

A polyhedral form simply means a solid bounded by flat planes (we call these flat planes CRYSTAL FACES). "A chemical compound" tells us that all minerals are chemicals, just formed by and found in nature.

The last half of the definition tells us that a crystal normally forms during the change of matter from liquid or gas to the solid state. In the liquid and gaseous state of any compound, the atomic forces that bind the mass together in the solid state are not present.

Therefore, we must first crystallize the compound before we can study it's geometry. Liquids and gases take on the shape of their container, solids take on one of several regular geometric forms. These forms may be subdivided, using geometry, into six systems.

But before we can begin to discuss the individual systems and their variations, let's address several other topics which we will use to describe the crystal systems. There's also some laws and rules we must learn. Way back in 1669, Nicholas Steno, a Danish physician and natural scientist, discovered one of these laws.

By examination of numerous specimens of the same mineral, he found that, when measured at the same temperature, the angles between similar crystal faces remain constant regardless of the size or the shape of the crystal. So whether the crystal grew under ideal conditions or not, if you compare the angles between corresponding faces on various crystals of the same mineral, the angle remains the same. Although he did not know why this was true (x-rays had not yet been discovered, much less x-ray diffraction invented), we now know that this is so because studies of the atomic structure of any mineral proves that the structure remains within a close set of given limits or geometric relationships.

If it doesn't, then by the modern definition of a mineral, we are not comparing the same two minerals. We might be comparing polymorphs, but certainly not the same mineral! (Polymorphs being minerals with the same chemistry, like diamond and graphite or sphalerite and wurtzite, but having differing atomic structure and, therefore, crystallizing in different crystal systems) Steno's law is called the CONSTANCY OF INTERFACIAL ANGLES and, like other laws of physics and chemistry, we just can't get away from it.

Now, some of you may be thinking: I have a mineral crystal that does not match the pictures in the mineral books. What you may have is a distorted crystal form where some faces may be extremely subordinate or even missing.

Distorted crystals are common and result from less-than-ideal growth conditions or even breakage and recrystallization of the mineral. However, remember that we must also be comparing the angles between similar faces.

If the faces are not present, then you cannot compare them. With many crystals we are dealing with a final shape determined by forces other than those of the interatomic bonding.

During the process of crystallization in the proper environment, crystals assume various geometric shapes dependent on the ordering of their atomic structure and the physical and chemical conditions under which they grow. If there is a predominant direction or plane in which the mineral forms, different habits prevail. Thus, galena often forms equate shapes (cubes or octahedrons), quartz typically is prismatic, and barite tabular.

To discuss the six crystal systems, we have to establish some understanding of solid geometry. To do this, we will define and describe what are called CRYSTALLOGRAPHIC

AXES. Since we are dealing with 3 dimensions, we must have 3 axes and, for the initial discussion, let's make them all equal and at right angles to each other. This is the simplest case to consider. The axes pass through the centre of the crystal and, by using them, we can describe the intersection of any given face with these 3 axes.

Mineralogists had to decide what to call each of these axes and what their orientation in each crystal was so that everyone was talking the same language. Many different systems arose in the early literature.

Then, as certain systems were found to have problems, some were abandoned until we arrived at the notational system used today. There exists two of these presently in use, complimentary to each other. One uses number notation to indicate forms or individual faces and the other uses letters to indicate forms. But let's get back to our 3 axes and we'll discuss these two systems later.

We are going to draw each axis on a sheet of paper and describe its orientation. All you need for this exercise is a pencil and paper. Make the first axis vertical, and we'll call it the c axis.

The top is the + end and the bottom is the - end. The second axis, the b axis, is horizontal and passes through the centre of the c axis. It is the same length as the c axis. The right end is the +, and the left is the -. The third axis is the a axis and passes at a right angle through the join of the b and c axes.

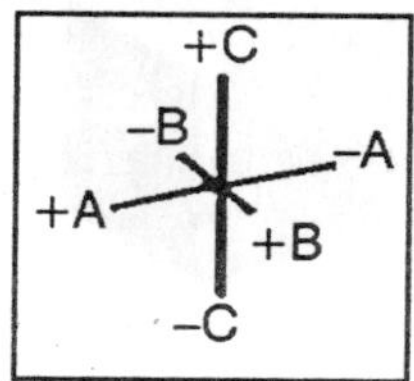

It is somewhat tricky to draw because, even though the a axis is the same length as the c and b axes, because it goes from front to back it appears shorter. It is hard to represent a 3-dimensional figure on the 2-dimensional surface of paper, but you can do it. You have to use a sense of perspective, as an artist would say.

The front end of a, which appears to come out of the paper, is the + and the back end the - (appears to be in the background or behind the paper).

This all sounds complicated, but look at Figure if you are having problems with drawing the final axis. We always refer to the axes in the order - a, b, c in any type of notation. The point of intersection of the three axes is called the AXIAL CROSS. Perspective is a key to drawing 3 dimensional objects on a flat 2D piece of paper.

Perspective is what makes railroad tracks look like they come together in the distance. It is also what causes optical illusions when trying to draw axial crosses, or line drawings of crystal models.

Perhaps you've looked at these lines and tried to decide which one comes forward, and which one recedes, and then have the illusion flip-flop so that it looks the other way. That's why they are labeled + and -. It can be confusing, so don't feel like the lone ranger.

We have now reached the point in our discussion that we can actually mention the six large groups or crystal systems that all crystal forms may be placed in. We will use our crystallographic axes which we just discussed to subdivide all known minerals into these systems.

The systems are:

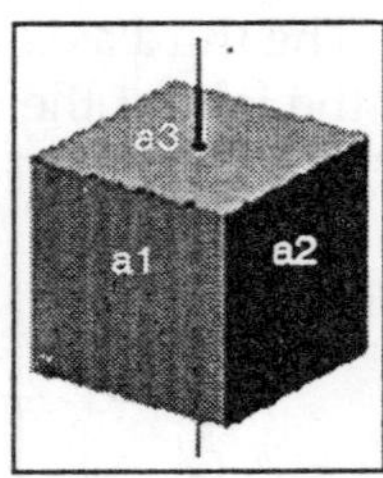

Cubic (aka Isometric

The three crystallographic axes are all equal in length and intersect at right angles (90 degrees) to each other. This is exactly what you drew to obtain. However, we now will rename the axes a1, a2, and a3 because they are the same length (a becomes a1, b becomes a2, and c becomes a3).

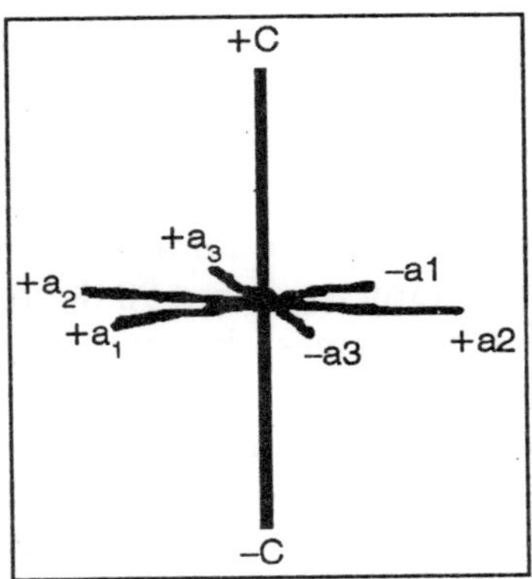

Tetragonal

Three axes, all at right angles, two of which are equal in length (a and b) and one (c) which is different in length (shorter or longer). If c was equal in length to a or b, then we would be in the cubic system.

Orthorhombic

Three axes, all at right angles, and all three of different lengths. Note: If any axis was of equal length to any other, then we would be in the tragonal system!

Hexagonal

Four axes! We must define this situation since it can not be derived from our . Three of the axes fall in the same plane and intersect at the axial cross at 120 degrees between the positive ends. These 3 axes, labeled a1, a2, and a3, are the same length. The fourth axis, termed c, may be longer or shorter than the a axes set. The c axis also passes through the intersection of the a axes set at right angle to the plane formed by the a set.

Monoclinic

Three axes, all unequal in length, two of which (a and c) intersect at an oblique angle (not 90 degrees), the third axis (b) is perpendicular to the other two axes. Note: If **a** and c crossed at 90 degrees, then we would be in the orthorhombic system!

Triclinic

The three axes are all unequal in length and intersect at three different angles (any angle but 90 degrees). Note: If any two axes crossed at 90 degrees, then we would be describing a monoclinic crystal.

As was stated earlier, all known crystal forms fit into the above six crystal systems. But why don't all crystals in a given set look the same? Or, stated differently, why can't I learn six crystal shapes and know all I need to know? Well, crystals, even of the same mineral, have differing CRYSTAL FORMS, depending upon their conditions of growth.

Whether they grew rapidly or slowly, under constant or fluctuating conditions of temperature and pressure, or from highly variable or remarkably uniform fluids or melts, all these factors have their influence on the resultant crystal shapes, even when not considering other controls.

We must touch on a type of notation often seen in mineral literature known as Miller Indices. Before William H. Miller (1801-1880) devised this mathematical system for describing any crystal face or group of similar faces (forms), there existed a considerable amount of confusion due to the many different descriptive systems.

Some of these systems used letter symbols to denote crystal faces and forms. Also different mineralogical "schools" existed as to how a given crystal should be viewed or oriented before assigning the crystallographic axes and then describing the various faces and forms present. If you were of the German school, you had one view; from the English school another thought; from the French school, still another opinion.

So the problem was really one of how to bring order to the literature's chaos. To the problem, Miller (University of Cambridge) applied relatively simple mathematics - the Universal Language.

To the lettering systems, Miller described the a,b,c (for hexagonal crystals his notation is four numbers long) intercepts of each planar crystal form as numbers and also made note of the form letter. His numbering system became widely accepted and is known as Miller indices. The numbers are presented as whole numbers (fractions are not allowed) and are the reciprocal of the actual intercept number, all whole numbers being reduced by their lowest common denominator. Here's a couple of simple examples from the cubic system.

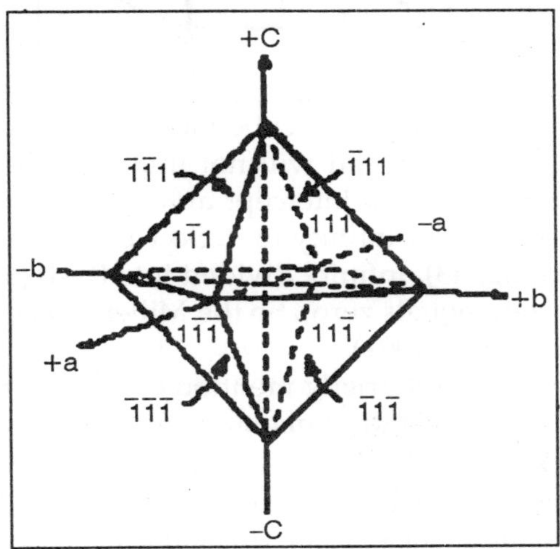

Let us first describe a face of an octahedron and later a cube using Miller's indices. First, we should realise that an octahedron is an eight-sided crystal form that is the simple repetition of an equilateral triangle about our 3 crystallographic axes.

The triangle is oriented so that it crosses the a1, a2, and a3 axes all at the same distance from the axial cross. This unit distance is given as 1.

Dividing 1 into the whole number 1 (it's a reciprocal, remember?) yields a value of 1 for each Miller number. So the Miller indices is (111) for the face that intercepts the positive end of each of the 3 axes. See Figure for all possible numbers for the 8 faces. Note: A bar over the number tells me that the intercept was across the negative end of the particular crystallographic axis. The octahedral form is given the letter designation of"o".

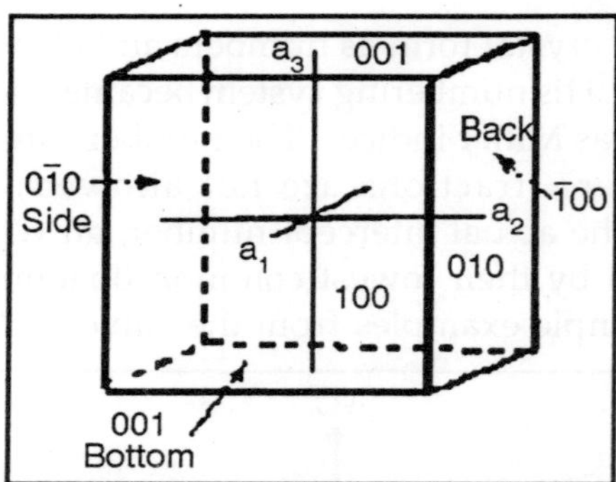

Now to the cube face. A cube face that intercepts the a3 (vertical) axis on the + end will not intercept the a1 and a2 axes. If the face does not intercept an axis, then we assign a mathematical value of infinity to it. So we start with Infinity, Infinity, 1 (a1, a2, a3). Infinity divided into 0 = 0 (any number divided into zero equals zero). So the Miller indices of the +a3 intercept face equals (001).

I think we should briefly mention cleavage at this point. CLEAVAGE is the preferred planar direction of breakage that many minerals possess. It is due to planes of weakness that exist in some minerals because the bonding strength between the atoms or molecules is not the same in every direction. Because crystals are composed of orderly arrangements of atoms or molecules, we really should expect cleavage to be present in many crystals.

The notation that denotes cleavage is derived in much the same manner as Miller indices, but is expressed in braces. So a cubic crystal, say diamond, no matter whether it exhibits cubic or octahedral crystal form, has an octahedral cleavage form that is given as.

Note: The Miller indices when used as face symbols are enclosed in parentheses, as the face for example. Form symbols are enclosed in braces, as the form for example. Zone symbols are enclosed in brackets, for example and denote a zone axis in the crystal. So in the discussion of cleavage (above), you must use braces to denote cleavage. Cleavage is analogous to form as cubic, octahedral, or pinacoidal cleavage and does not refer to just one face of a form.

Now we are ready to discuss ELEMENTS OF SYMMETRY. These include PLANES OF SYMMETRY, AXES OF SYMMETRY, and CENTRE OF SYMMETRY. These symmetry elements may be or may not be combined in the same crystal. Indeed, we will find that one crystal class or system has only one of these elements!

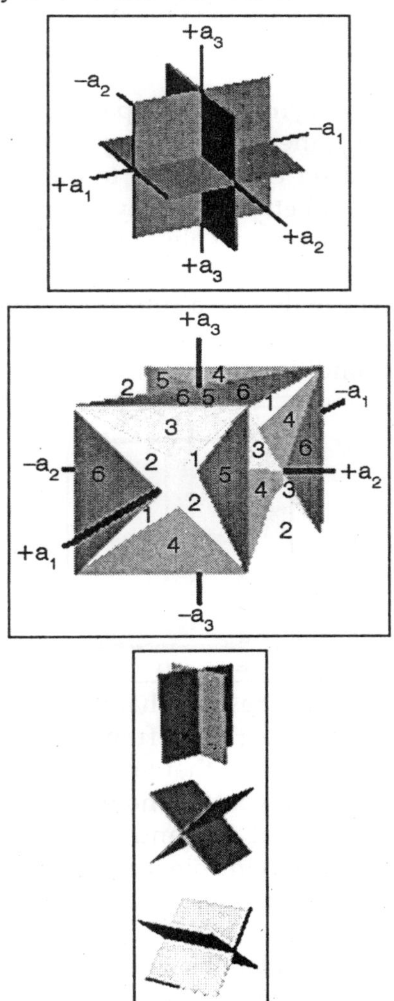

Huh? These parts, when put together, make the planes in

. Any two dimensional surface (we can call it flat) that, when passed through the centre of the crystal, divides it into two symmetrical parts that are MIRROR IMAGES is a PLANE OF SYMMETRY. I repeat: any plane of symmetry divides the crystal form into two mirror images. Planes of symmetry are often referred to as mirror image planes. Let's discuss a cube again. A cube has 9 planes of symmetry, 3 of one set and 6 of another. We must use two figures to easily recognize all of them.

In the planes of symmetry are parallel to the faces of the cube form, in the planes of symmetry join the opposite cube edges. The second set corresponds to the octahedral crystal form. Planes of symmetry are always possible crystal forms. This means that, although not always present on many natural crystals, there exists the possibility that other crystal faces may be expressed. So even though a cube form does not present an octahedral face, it is always possible that it could have formed under the right conditions.

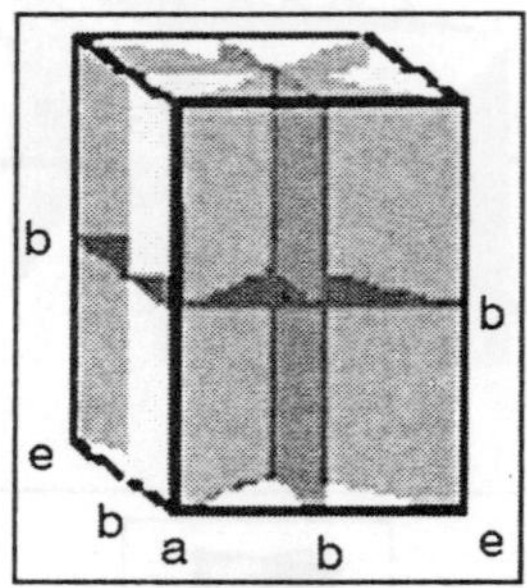

The typical human has two hands, right and left. Place them together palms facing away from you and the tips of the thumbs touching. Assuming that you have the same number of fingers on each hand, you will note that your right hand is the mirror image of your left and vise versa. The average person is symmetrical, having binary symmetry vertically from the head to the feet when viewed from the front or back (bilateral symmetry).

You can have a lot of laughs with friends and a long mirror using this symmetry element. You need at least one other person to do this so you both can view the results! Using as guides, take a wood or plastic cube and see if you can draw with a marker all the planes of symmetry that are present.

It is sometimes convenient to designate planes of symmetry as axial, diagonal, principle, or intermediate. is an example of the 5 planes of symmetry of the tetragonal system and the proper abbreviated notation.

AXES OF SYMMETRY can be rather confusing at first, but let's have a go at them anyway. Any line through the centre of the crystal around which the crystal may be rotated so that after a definite angular revolution the crystal form appears the same as before is termed an axis of symmetry. Depending on the amount or degrees of rotation necessary, four types of axes of symmetry are possible when you are considering crystallography (some textbooks list five). Given below are all possible rotational axes:

When rotation repeats form every 60 degrees, then we have sixfold or HEXAGONAL SYMMETRY. A filled hexagon symbol is noted on the rotational axis.

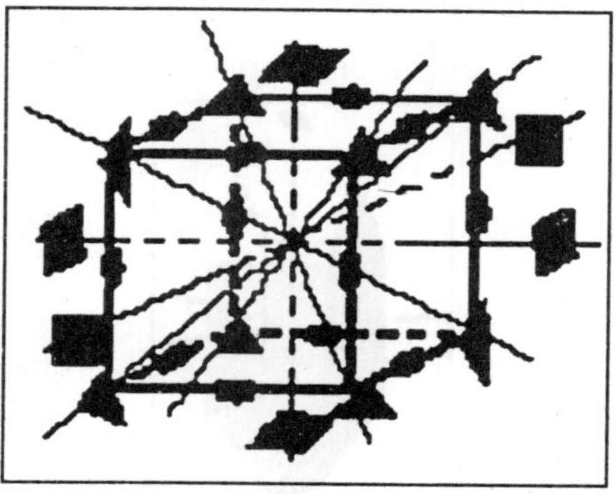

When rotation repeats form every 90 degrees, then we have fourfold or TETRAGONAL SYMMETRY. A filled square is noted on the rotational axis.

When rotation repeats form every 120 degrees, then we have threefold or TRIGONAL SYMMETRY. A filled equilateral triangle is noted on the rotational axis.

When rotation repeats form every 180 degrees, then we have twofold or BINARY SYMMETRY. A filled oval is noted on the rotational axis.

When rotation repeats form every 360 degrees, then we use a filled circle as notation. This one I consider optional to list as almost any object has this symmetry. If you really want to know the truth, this means NO SYMMETRY!!

Note that rotational axes may be on the plane of the face, on the edge of where two faces meet, or on the point of conjunction of three or more faces. On a complete crystal form, the axis must pass through the centre of the crystal and exist at the equivalent site on the opposite side of the crystal as it entered. Take a solid cube, made of wood or plastic (a clear plastic cube box works well for this exercise). Mark, using the rotational notation, every four-, three-, and two-fold axis of rotation that you can find. I think you will be surprised how many there are! Examine (the cube from hades!) to see how many symbols you can draw on your cube.

Now I'm sorry that this is not all there is to rotational axes, but there is another situation that we must consider -- AXES OF ROTARY INVERSION. This is where the twisted mind has one up on the rest of us (there's a pun in there somewhere!). We will consider a couple of simple examples.

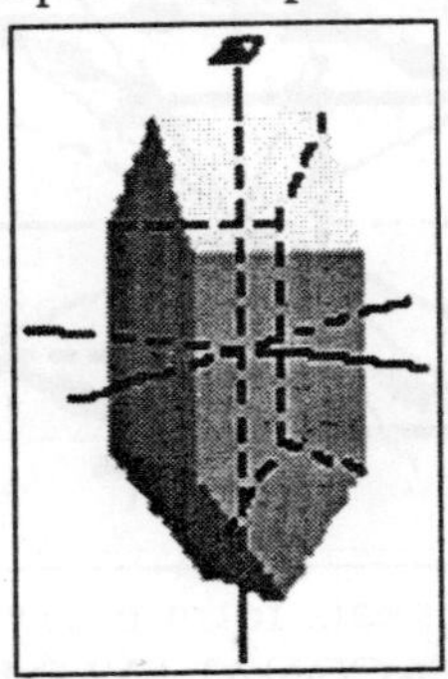

First, let's examine a a crystal as drawn in a at left. Use a piece of"2 by 2" board and make this crystal form by cutting off the ends so the wood block looks like the drawing. Hold the block in your left hand with your thumb on the top and in the centre of the 2-face edge join (long axis) and your index finger on the same join on the underside. Your palm will be toward your body. Align your two fingers so that you are looking straight down on your thumb and can not see end of your index finger. The top of the block will appear as 2 equal-sized faces, sloping away from you. If you rotate the block 180 degrees, the faces will be appear back in the same position (2-fold axis of rotation), but here's the tricky part -- rotate the specimen 90 degrees and then turn your wrist where your index finger is on top (easiest done by turning your wrist counterclockwise). You will see that the block's faces appear in the original position in the original position.

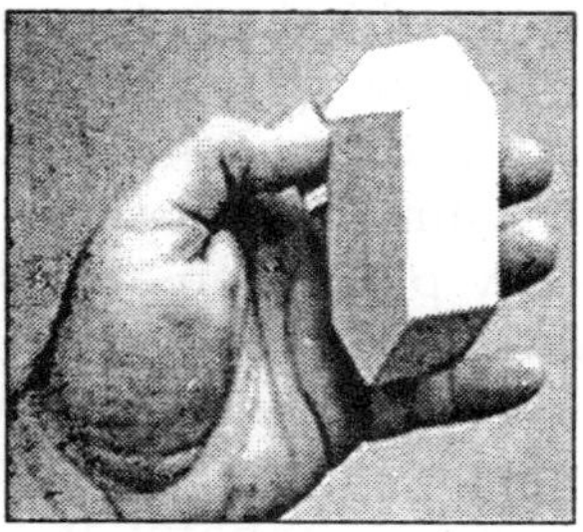

Fig. Wooden or Plastic Models are Known in a Mineralogy Class as'Idiot Blocks'

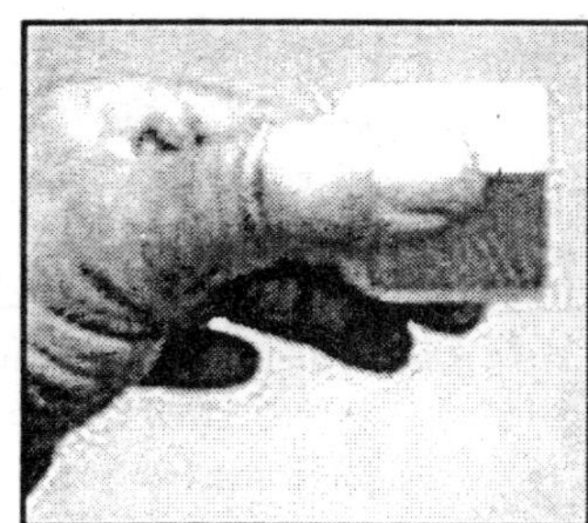

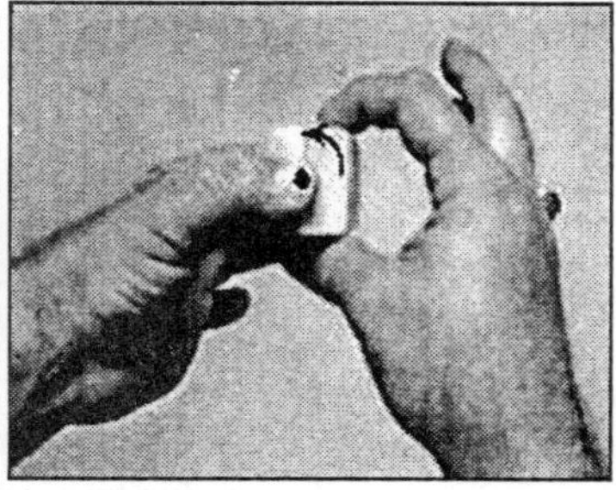

Fig: Block Rotated 90 Degrees Around the Axis Shown by the Dot

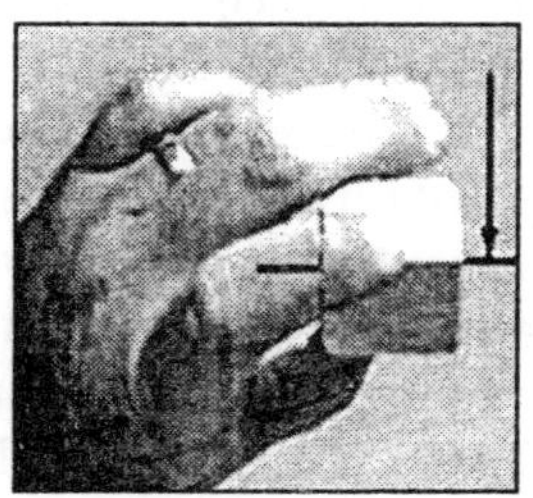

Fig. Block Rotated Counterclockwise 180°
on the Axis Shown by the Arrow.

See the series of photos (Figures) if you get confused. Some textbooks term these axes rotary reflection axes or rotoinversion axes. There may be 1-, 2-, 3-, 4-, and 6-fold rotary inversion axes present in natural crystal forms, depending upon the crystal system we are discussing. I refer you to Klein and Hurlbut's Manual of Mineralogy (after J. S. Dana) if you want to sharpen your axes of rotary inversion skills. With axes of rotation, there is a graphical notation used which looks like a very bold type-face comma. For axes of rotary inversion, the same symbol is used, but appears dashed.

Both types of symmetrical rotational axes (discussed above) are commonly plotted on a circle (representing the complete cycle of one 360 degree rotation). The simple axes of rotation symbol for a face is plotted at the centre of the circle and the axes of rotation and rotary inversion are plotted on the circle's boundary at whatever rotational angle is appropriate for examples.

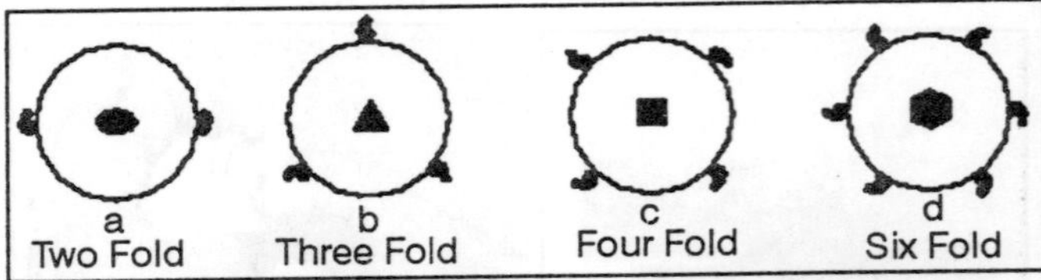

We have finally come to our last topic of geometric crystallography -- the CENTRE OF SYMMETRY. Most crystals have a centre of symmetry, even though they may not possess either planes of symmetry or axes of symmetry. Triclinic crystals usually only have a centre of symmetry. If you can pass an imaginary line from the surface of a crystal face through the centre of the crystal (the axial cross) and it intersects a similar point on a face equidistance from the centre, then the crystal has a centre of symmetry. We may discuss this in a little more detail in the article about the triclinic system.

We now have to consider the relation of geometrical symmetry to CRYSTALLOGRAPHIC SYMMETRY. The crystal face arrangement symmetry of any given crystal is simply an expression of the internal atomic structure. This internal structure is generally alike in any parallel direction. But we must keep in mind that the relative size of a given face is of no importance, only the angular relationship or position to other given crystal faces.

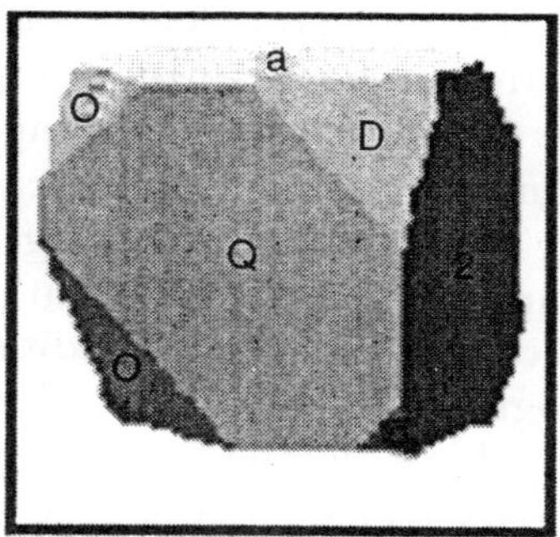

Let's consider a crystal in the cubic system with both cube {001} and octahedral forms represented . In our figure, we have used the letter designation of -a- for the cube faces and -o- for the octahedral faces. Despite the initial observation that both the various cube and octahedral faces are unequal in size, the example displays all the symmetry elements and relationships of a crystal from the cubic system. I hope you now begin to grasp the difficulty of learning crystallography using natural crystals. Due to a variety of factors, many natural crystals have some degree of distortion to their growth, causing the faces to vary in size and sometimes shape. In college mineralogy, this problem was resolved by requiring the classroom use of a set of crystal forms, sometimes made of wood or plastic. These sets were not-so-fondly termed"idiot blocks" by exasperated students. Once you mastered the various forms and understood symmetry planes, rotational axes, and form names, then you became recognized as a"complete idiot" and could go on to examine real minerals!

Depending upon what elements of symmetry are present, all crystals may be divided into 32 distinct groups called CLASSES OF SYMMETRY. Remember, we concern ourselves with the symmetry elements we learned above, not the malformed crystal shapes of most minerals. Only forms which belong to the same class can occur in combination together in nature.

We can not find a cube face on a hexagonal crystal. Likewise, we will never discover the rhombic dipyramid termination of a hexagonal crystal on a tetragonal crystal. So our laws, rules, and symmetry elements previously discussed prevent chaos in our beautifully symmetrical world of crystallography. Certainly, when dealing with real crystals, distortion problems can arise! Think of capillary pyrite. Here you have a cubic crystal which, due to a growth phenomenon, has one axis nearing infinity in length in relation to the other two. But this is caused by special conditions during growth, not the crystallography.

There are graphical methods of plotting all possible crystal faces and symmetry elements on a type of diagram called a stereo net. Stereo nets give a way to represent three-dimensional data on a two-dimensional surface (a flat sheet of paper). A discussion of stereo nets is out of the scope of this paper because to present it adequately would require much graphing, mathematics, and that each reader possess stereo-net paper. If you wish to attempt any exercises with stereo nets, I refer you to the previously mentioned textbooks. You can purchase the graph paper at most major college and university bookstores.

Well, if your faces are all shining, your symmetry now in order, and your axes properly aligned, then stay tuned for the next article when we consider crystal forms and the 32 symmetry classes. Then we will have the background necessary to discuss the six crystal systems!

ROCK FORMING MINERALS

ROCKS AND MINERALS

It is not easy to tell the difference between rocks and minerals because there are so many kinds of them. It takes years of study to be able to accurately identify a mystery rock and even then rockhounds want to know where the specimen came from.

All rocks are made of 2 or more minerals, but minerals are not made of rocks.

Rock Words

There are many common names for rocks and the usually give you an idea of how big the rock is. Here are a few:

- Mountain - huge, giant hunk of rock that is still attached to the earth's crust, doesn't move, tall
- Boulder - large, taller than a person
- Rock - large, you could get your arms around it or a bit smaller but it is usually jagged,
  broken off a bigger piece of rock
- River rock - round rocks that are along the edge and at the bottom of fast-flowing rivers
- Stone - medium, you could hold it in two hands
- Pebble - small, you can hold it with two fingers, could get stuck in your shoe, usually rounded
- Sand - made up of tiny pieces of rock, grains of sand
- Grain - tiny, like a grain of rice or smaller, often found on a beach
- Dust - really fine powder that is mixed in with sand or soil
- Speck - as in a speck of dirt

Minerals

- A mineral is the same all the way through. That is one reason we speak of
  a sample or a specimen rather than a rock.
- There are about 3000 known minerals on earth.

ROCKS AND MINERALS FORMATION

Attention Grade 4 Ontario teachers in the Greater Toronto Area: Book a school visit for your rocks and minerals unit with Rock Talks by Rockhounds!

The Earth's Crust

The whole earth is made of rocks and minerals. Inside the earth there is a liquid core of molten rock and on the outside there is a hard crust. If you compare the earth to an egg, the shell on an egg is like the crust on the earth.

The crust is made up of rocks and minerals. Much of the crust is covered by water, sand, soil and ice. If you dig deep enough, you will always hit rocks.

Below the loose layer of soil, sand and crumbled rocks found on Earth is bedrock, which is a solid rock:

- The Crust makes up less than 1% of the Earth's mass (0.4%)
  It is made of oxygen, magnesium aluminum, silicon calcium, sodium potassium, iron.
  There are 8 elements that make up 99% of the Earth's crust.
  The continents are about 35 km thick and the ocean floors are about 7 lm thick.
- The Mantle is the solid casing of the Earth and is about 2900 km thick.
  It makes up about 70% of the Earth's mass (68.1%).
  It is made up of silicon, oxygen, aluminum and iron.
- The Core is mainly made of iron and nickel and makes up about 30% of the Earth's mass (31.5%).
  The Outer Core is 2200 km thick and is liquid and the Inner core is 1270 km thick and is solid.

Rocks

The rocks you see around you - the mountains, canyons and riverbeds, are all made of minerals. A rock is made up of 2 or more minerals.

- Think of a chocolate chip cookie as a rock.
- The cookie is made of flour, butter, sugar and

chocolate.

- The cookie is like a rock and the flour, butter, sugar and chocolate are like minerals.
- You need minerals to make rocks, but you don't need rocks to make minerals.
- All rocks are made of minerals.

Minerals

A mineral is composed of the same substance throughout.

If you were to cut a mineral sample, it would look the same throughout. There are about 3000 different minerals in the world. Minerals are made of chemicals - either a single chemical or a combination of chemicals. There are 103 known chemical elements. Minerals are sorted into 8 groups.

Some common examples have been listed for each:

- Native Elements ~ copper, silver, gold, nickel-iron, graphite, diamond
- Sulfides ~ sphalerite, chalcopyrite, galena, pyrite
- Halides ~ halite, fluorite
- Oxides and Hydroxides ~ corundum, hematite
- Nitrates, Carbonates, Borates ~ calcite, dolomite, malachite
- Sulfates, Chromates, Molybdates, Tungstates ~ celestite, barite, gypsum
- Phosphates, Arsenates, Vanadates ~ apatite, turquoise
- Silicates ~ quartz, almandine garnet, topaz, jadeite, talc, biotite mica

Crystals

Crystals are minerals that have had the chance to grow in the shape that they were meant to be. Just like your DNA determines the colour of your eyes, how tall you will get to be and the shape of your bones, the chemicals that a mineral is made of determines what shape it gets to be. We can tell different minerals apart by what crystal shape they are.Sometimes minerals form in spaces where there is not a lot of room, so they don't have a crystal shape. When there is just a big hunk of a mineral, it is called a massive mineral. If there is a definite shape with easy to see flat sides, it is called a mineral crystal.

Most of the earth's crystals were formed millions of years ago. Crystals form when the liquid rock from inside the earth cool and harden. Sometimes crystals form when liquids underground find their way into cracks and slowly deposit minerals. Most mineral crystals take thousands of years to "grow" but some like salt (halite) can form so quickly that you can watch them grow at home!Some people think of crystals as clear pretty rocks that are used for jewelry. Amethyst is a very common quartz crystal. Crystals do not have to be clear, but those are the kinds you will usually see in the stores.

Soil, Sand and Dirt

When rocks break down into smaller and smaller pieces, they turn into sand. If you look at the sand under a microscope, you will see that sand is made up of the same minerals as the rocks that the sand came from. When plants start to sprout up in sand, it is turning from being just small bits of rock to being soil.

Soil is very important to life on earth. It supports plant life. We could not live without plants. Soil is made up of sand and decomposing plants and animals. Soil has many names including: clay, silt, mud, dirt, topsoil, dust, potting soil and humus

The Rock Cycle

Rocks are constantly being formed, worn down and then formed again. This is known as the Rock Cycle. It is like the water cycle but it takes a lot longer. It takes thousands and millions of years for rocks to change.

TYPES OF ROCKS

Rocks are divided into 3 Types. They are classified by how they were formed:

- Igneous
- Sedimentary
- Metamorphic

Some good sites that help explain this are:

- The Stupid Page of Rocks Describes metamorphic, sedimentary and igneous rocks
- Rocks and the Rock Cycle A well written description of the Rock Cycle.

Igenous Rocks

Igneous means made from fire or heat. When volcanoes erupt and the liquid rock comes up to the earth's surface, then new igneous rock is made. When the rock is liquid and inside the earth, it is called magma.

When the magma gets hard inside the crust, it turns into granite. Most mountains are made of granite. It cools very slowly and is very hard. When the magma gets up to the surface and flows out, like what happens when a volcano erupts, then the liquid is called lava. Lava flows down the sides of the volcano.

When it cools and turns hard it is called obsidian, lava rock or pumice - depending on what it looks like:

- Igneous rocks form when molten lava (magma) cools and turn to solid rock.
  The magma comes from the Earth's core which is molten rock .
  The core makes up about 30% of the Total Earth Mass (31.5%)
- Obsidian is nature's glass. It forms when lava cools quickly on the surface. It is glassy and smooth.
- Pumice is full of air pockets that were trapped when the lava cooled when it frothed out onto the surface. It is the only rock that floats.

There are 5 kinds of igneous rocks, depending on the mix of minerals in the rocks.

- Granite contains quartz, feldspar and mica
- Diorite contains feldspar and one or more dark mineral. Feldspar is dominant.
- Gabbro contains feldspar and one or more dark mineral. The dark minerals are dominant.
- Periodotite contains iron and is black or dark.

- Pegmatite is a coarse-grained granite with large crystals of quartz, feldspar and mica.

Sedimentary Rocks

When mountains are first formed, they are tall and jagged like the Rocky Mountains on the west coast of North America. Over time (millions of years) mountains become old mountains like the Appalachian Mountains on the east coast of Canada and the United States. When mountains are old, they are rounded and much lower. What happens in the meantime is that lots of rock gets worn away due to erosion. Rain, freeze/thaw cycle, wind and running water cause the big mountains to crumble a little bit at a time.

Eventually most of the broken bits of the rock end up in the streams and rivers that flow down from the mountains. These little bits of rock and sand are called sediments. When the water slows down enough, these sediments settle to the bottom of the lake or oceans they run into. Over many years, layers of different rock bits settle at the bottom of lakes and oceans.

But a stack of telephone books is very heavy and would squish anything that was underneath. Over time the layers of sand and mud at the bottom of lakes and oceans turned into rocks. These are called sedimentary rocks.

Some examples of sedimentary rocks are sandstone and shale. Sedimentary rocks often have fossils in them. Plants and animals that have died get covered up by new layers of sediment and are turned into stone.

Most of the fossils we find are of plants and animals that lived in the sea. They just settled to the bottom. Other plants and animals died in swamps, marshes or at the edge of lakes. They were covered with sediments when the size of the lake got bigger.

When large amounts of plants are deposited in sedimentary rocks, then they turn into carbon. This gives us our coal, oil, natural gas and petroleum. A large sea once covered the central part of Canada and the climate was very tropical.

In time, sedimentary rocks formed there. That is why we find dinosaur fossils in Alberta and the area is a good source of natural fuels.

- Sedimentary rocks cover 75% of the earth's surface. Most of the rocks found on the Earth's surface is sedimentary even though sedimentary rocks only make up less than 5% of all the rocks that make up Earth.
- When rocks are exposed to the elements – air, rain, sun, freeze/thaw cycle, plants – erosion occurs and the little bits of rock worn away get deposited as sediments.
  Over time, these sediments harden as they get buried by more sediments and turn into sedimentary rocks.
- Sedimentary rocks are usually formed in layers called strata.
- There are 6 main kinds of sedimentary rocks depending on the appearance of the rock.
  - Conglomerate rock has rounded rocks (pebbles, boulders) cemented together in a matrix.
  - Sandstone is a soft stone that is made when sand grains cement together. Sometimes the sandstone is deposited in layers of different coloured sand.
  - Shale is clay that has been hardened and turned into rock. It often breaks apart in large flat sections.
  - Limestone is a rock that contains many fossils and is made of calcium carbonate and/or microscopic shells.
  - Gypsum, common salt or Epsom salt is found where sea water precipitates the salt as the water evaporates.
  - Breccia has jagged bits of rock cemented together in a matrix.

Metamorphic Rocks

Metamorphic rocks are rocks that have changed. The

word comes from the Greek "meta" and "morph" which means to change form. Metamorphic rocks were originally igneous or sedimentary, but due to movement of the earth's crust, were changed.

If you squeeze your hands together very hard, you will feel heat and pressure.

When the earth's crust moves, it causes rocks to get squeezed so hard that the heat causes the rock to change. Marble is an example of a sedimentary rock that has been changed into a metamorphic rock.

- Metamorphic rocks are the least common of the 3 kinds of rocks.
  Metamorphic rocks are igneous or sedimentary rocks that have been transformed by great heat or pressure.
- Foliated metamorphic rocks have layers, or banding.
  - Slate is transformed shale. It splits into smooth slabs.
  - Schist is the most common metamorphic rock. Mica is the most common mineral.
  - Gneiss has a streaky look because of alternating layers of minerals.
- Non-foliated metamorphic rocks are not layered.
  - Marble is transformed limestone.

Quartzite is very hard.

EROSION

Erosion is a key part of the Rock Cycle. It is responsible for forming much of the interesting landscape that is around us. It is also a major problem as people live in areas in large numbers and get used to the environment being in a certain way. People can do things to increase erosion or slow it down.

Erosion happens mainly as a result of weathering – the effect of water, temperature and wind on the landscape.

Water causes much erosion. When it falls as acid rain, it can dissolve rocks that are sensitive to acid. Marble and limestone weather when exposed to the rain.

When the rain falls very heavily, as in monsoons, then flooding can happen.

Fig. Rivers with a Lot or Rushing Water can Caus Mud Slides and Erode River Banks.

The action of waves on a beach causes much erosion. The waves pound on the rocks and over time, cliffs crumble. That is why you will often find sand and little pebbles on beaches. Rushing water, like what you find in rivers that move quickly in the mountains or strong waves on the shores of oceans, roll rocks around. This causes the sharp edges of the rocks to get knocked off and that is why river rocks are so smooth and beach pebbles look polished.

- *Acid Rain*: Chemicals in the air combine with precipitation
  When it rains it dissolves certain minerals sensitive to acid.
- *Leaching by ground water*: Water soaks into the soil, picks up chemicals
  This allows the water to leach or dissolve rocks it comes in contact with at bedrock.
- *Wave action at the beach*: The waves tumble rocks
  Rocks get ground down by the sand particles already

on the beach, rocks smash against each other and break.

- *Fast moving water*: Rocks get picked up and carried when water runs swiftly

  By bouncing along a river and smashing into other rocks, the sharp edges get knocked off.
- *Glaciers*: Large sheets of ice pick up large rocks, scrape bedrock

  Rocks tumble in under-glacier rivers when glaciers melt.
- *Precipitation/Floods*: Heavy rain can cause floods which move and break rocks

The freeze/thaw cycle causes mountains to crumble over time and large rocks to break down into little rocks. When water gets into cracks in the rocks, this water expands during the freeze cycle, making the cracks bigger.

Then when the cracks fill up with water in the thaw period. This allows more water to go deeper into the rock which will make the rocks split apart when they freeze again. The power of frozen water expanding can be seen when you leave.

A glass bottle filled with liquid in the freezer:

- Wind, when it carries bits of sand and grit, can blast away layers of rocks.
- The wind can easily pick up little bits of sand and then sandblast the rocks that are in the wind's way.
- Sometimes only the soft layers of the rock are eroded, leaving interesting shapes.
- This kind of erosion usually only happens in very dry, desert like areas.
- Other causes for rocks to break down and erode:
  - How hard/tough mineral is: softer, more friable rocks and minerals break up easily
  - Plant roots growing: plants get nutrients from the soil, seek out certain minerals like potash,

    apatite for fertilizer, small roots go in cracks and break up mineral or rock when the root grows bigger

- Rock Falls: rocks tumbling down from a cliff or steep mountainside cause rocks to break up

Contact with soil: certain soils have chemicals in them that react with the

ROCK-FORMING MINERALS

Of the simple undecomposable substances which chemists call elements, and of which rather more than seventy have been discovered, only sixteen enter at all largely into the composition of the earth's crust, so far as this is accessible to observation. It is estimated that 98% of the crust is made up of the following eight elements, arranged in the order of abundance, with the per centages as calculated by F. W. Clarke.

Oxygen........ ....47.07
Silicon....... . . . . . 28.06
Aluminium......... 7.90
Iron............ .. 4.43
Calcium............ 3.44
Magnesium........2.40
Sodium.............2.43
Potassium...........2.45

The remaining eight elements, titanium, carbon, sulphur, hydrogen, chlorine, phosphorus, manganese, and barium, are far less abundant, but still of considerable importance.

Only two of these elements, carbon and sulphur, are found in a more or less impure state as minerals or rock masses; the others occur as compounds, formed by the union of two or more of them.

A mineral is a natural, inorganic substance, which has a homogeneous structure, definite chemical composition and physical properties, and usually a definite crystal form.

Crystals are solids of more or less regular and symmetrical shape, bounded, usually, by plane surfaces. The number of known crystal forms is very great, and yet they may be all grouped in six systems, which are characterized by the relations of their axes. The axes of a crystal are imaginary lines, which connect the centres of opposite faces, or opposite edges, or opposite solid angles, and which intersect one another at a point in the interior of the crystal.

The Systems of Crystal Forms have received many names, the following being those which are most generally used in this country: -

Isometric System (Monometric, Cubical, Regular)

In this system the three axes are of equal length and intersect one another at right angles.

Tetragonal System (Dimetric, Pyramidal)

The axes intersect at right angles, but while the lateral axes are of equal length, the vertical axis is longer or shorter than the laterals.

Hexagonal System

Here four axes are employed, three equal lateral axes intersecting at angles of 60°, and a vertical axis, which is perpendicular to and longer or shorter than the laterals.

Orthorhombic System (Rhombic, Trimetric)

The three axes intersect at right angles and are all of different lengths.

Monoclinic System (Monosymmetric, Oblique)

All three axes are of different lengths; the two laterals are at right angles to each other, while the third is oblique to one of the former.

Triclinic System (Anorthic, Asymmetric)

Three axes of unequal lengths and oblique to one another: It is important to bear in mind the relations which the forms sustain toward one another. For example, a regular octahedron may be derived from a cube by evenly paring off the eight solid angles, until the planes thus produced intersect one another, the centres of the faces of the cube becoming the apices of the solid angles of the octahedron. Conversely, a cube may be formed from an octahedron by symmetrically truncating the angles, until the planes thus formed intersect. By slicing away the twelve edges of a cube or an octahedron a dodecahedron will result. These crystal forms are, therefore, so related as to be all derivable one from another, and the relations of their axes remain unchanged; all three forms may be assumed by the same mineral, and they thus properly belong in the same system. Similar relations may be observed between the crystal forms of the other systems.

It might be supposed that the crystal systems and the relations of their imaginary axes were merely mathematical devices to reach a convenient classification of forms. Such a conclusion would, however, be a very erroneous one. Crystalline form is an expression of molecular structure, and the physical properties of minerals are closely related to their mathematical figure.

It is clear that these physical properties are not inherent in the molecules of the mineral, but are conditioned by the way in which the molecules are built up into the crystal. Amorphous substances refract light equally in all directions, and are thus called isotropic; but when an amorphous substance crystallizes, it assumes the qualities proper to its crystal form. Thus water is isotropic, while the hexagonal crystals of ice are singly refractive in only one direction, doubly refractive in all others. The same substance may, under different circumstances, crystallize in different systems, and will then display the properties appropriate to each system.

Not only the refractive powers of a crystal, but also its mode of expansion when heated, and its conductivity of electricity and heat are controlled by the molecular structure which determines its shape.

The crystals of the isometric system, which have their three axes of equal length, are singly refractive in all directions, expand equally when heated, and conduct heat and electricity equally in all directions. Those of the tetragonal and hexagonal systems, which have one axis longer or shorter than the others, are doubly refractive along the lateral axes, expand equally when heated, and show equal conductivity along these axes. Along the principal axis they are singly refractive, expand to a different degree when heated, and display a different conductivity along this axis than along the others. In the orthorhombic, monoclinic, and triclinic systems, which have all the axes of unequal lengths, the crystals are singly refractive in two directions; they expand unequally and conduct differently along all their axes.

The optical properties of minerals are of great value in the study of rocks, and by the aid of the polarizing microscope very minute crystals may be identified.

Most inorganic substances which are solid under any circumstances are capable of assuming a crystal form, so that solidification and crystallization are usually identical. For the formation of large and regular crystals, it is necessary that the process be gradual and that space be given for the individual crystals to grow. Usually crystallization begins at many points simultaneously, and the crystals crowd upon one another, resulting in a mass of more or less irregular crystalline grains. The same substance which, when very rapidly solidified, forms an amorphous glass, will give rise to distinct crystals, if slowly solidified.

Crystallization requires that the molecules be free to move upon each other, and thus to arrange themselves in a definite fashion. It may take place either by the deposition of a solid from solution, by cooling from a state of fusion, or by solidification from the condition of vapour. In all cases the size and regularity of the crystals depend upon the time and space allowed for their growth. In a manner not yet understood, amorphous solids may be converted into crystalline aggregates. This has been observed in the case of certain glassy volcanic rocks, which, though amorphous when first solidified, have gradually become crystalline, without losing their solidity, and a similar change has been observed in certain artificial glasses. This process is called devitrification.

The actual steps of crystallization may be observed by slowly evaporating a solution of some crystalline salt under the microscope. The first visible step in the process is the appearance of innumerable dark points in the fluid, which rapidly grow, until their spherical shape is made apparent.

The globules then begin to move about rapidly and arrange themselves in straight lines, like strings of beads, and next suddenly coalesce into straight rods. The rods arrange themselves into layers, and thus build up the crystals so rapidly, that it is hardly possible to follow the steps of change. In certain glassy rocks, which solidified too quickly to allow crystallization to take place, the incipient stages of crystals, in the form of globules and hair-like rods, may be detected with the microscope.

Chapter 2

Rocks

ORIGIN

THE ARCHAEAN ROCKS

This is a problem which has given rise to a great deal of discussion, but a solution appears to be near. Independently, in many countries, observers have reached the conclusion that these rocks are divisible into two great series, a schist series composed chiefly of highly metamorphosed sedimentary and volcanic rocks, and a gneissoid granite series, which is intrusive and later than the former.

Assuming that this conclusion is true, at least as a working hypothesis, it involves certain curious consequences. Surface lava flows and volcanic tuffs, and still more, sedimentary rocks, necessarily imply a solid floor upon which they were laid down, but of this floor not a trace has anywhere been found. The question immediately arises, what has become of it? No answer to this question can yet be given, but apparently the most likely suggestion is that the ascending floods of molten magma, which gave rise to the gneissoid granites, must have melted and assimilated it.

If this were only a local phenomenon, there would be nothing very surprising about it, but it would seem to be true of the entire globe, and this is a startling conclusion. We are then to suppose that a solid crust, however formed, was for a very long time sufficiently rigid and stable to allow a great thickness of sedimentary and volcanic rocks to be accumulated upon it and then was ingulfed and destroyed by a universally ascending magma, though it is not necessary to suppose that this took place simultaneously over the whole earth, or even within a relatively short period of time; it may have required ages in the accomplishment.

Furthermore, it must not be forgotten that remnants of the floor may yet be discovered in little-known regions. If this complete and universal assimilation actually took place, it is an absolutely unique phenomenon in the recorded history of the earth, though something more or less similar may have happened many times before that record began.

Many other hypotheses have been propounded to account for the origin of the Archaean rocks, but as they are not supported by any strong evidence, it is not worth while to consider them here; several of them have been formally abandoned by their authors.

That the oldest known rocks were not the first to be formed is manifest from the derivative nature of many of them, for sediments necessarily imply some preexisting rock to furnish the materials, and volcanic outbursts involve a solid surface through which they break.

From the extreme degree of dynamic and thermal metamor-phism which the Archaean rocks have undergone, we should not expect to find any recognizable fossils in them. On the other hand, there are indirect evidences that life was already present on the earth at that period. The limestones, iron ores, and graphite found in these rocks appear to have been organically accumulated, but it is possible that they were chemically formed, and so the evidence, while probable, is not altogether conclusive.

ENVIRONMENTS OF ROCK ORIGIN

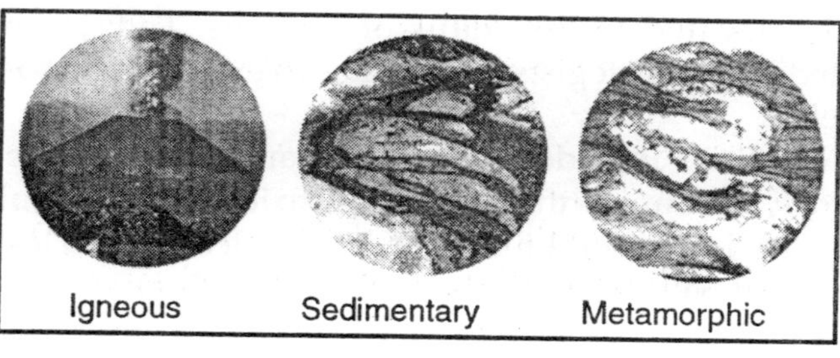

Geologists have defined three environments in which rocks form:

- *Igneous*: Igneous rocks form as molten rock cools and solidifies. Two environments are distinguished:
  - *Underground*: in which case the melt is called'magma' and the rock that results from its solidification is described as'intrusive'.
  - *On the surface*: in which case the melt is called'lava' and the rock that results is described as'extrusive' or'volcanic'.
- *Sedimentary*: Sedimentary rocks form as grains of sediment are attached to each other by a cement or by the interlocking of grains. The environment in which this occurs is at or close to the earth's surface and is characterized by relatively low temperatures and pressures.
- *Metamorphic*: Metamorphic rocks form as pre-existing rocks respond to a radical change in environment, most commonly an increase in temperature, pressure, and/or an infusion of hot, mineralized fluids.

Fact or Fiction

It is important to note that in general, these rock-forming processes and the environments in which they take place cannot be observed.

Solidification of magma, most changes in temperature and pressure sufficient to cause metamorphism, and most processes that attach sediment grains to one another occur out of sight beneath the earth's surface*. They are, therefore, not observed 'facts', but hypotheses that geologists in the 18th and 19th centuries constructed to explain features of rocks that they examined. Gradually, these hypotheses gained acceptance and became elevated to the status of theories. Today they are accepted by all geologists. Our task is to examine the basis for this belief.

Exceptions include the eruption and cooling of lava to form rock, the transformation of snow to ice, and the formation of salt rock in evaporating saline lakes, such as the Dead Sea or the Great Salt Lake, all of which can be observed.

STRUCTURE

ROCK STRUCTURE

Most rocks are not uniform throughout. On a scale usually best measured in millimetres or centimetres, they are composed of individual mineral grains that vary in size, shape and composition. The geometric characteristics of and relationships between these small-scale rock features constitute rock texture. Rock commonly also vary on larger scales, best measured in centimetres to metres to kilometres. The individual, contrasted, larger-scale features of rocks are called 'structures'. Our task will be to see if there are rock structures that can provide clues to a rock's formational environment: whether it's igneous, sedimentary or metamorphic.

There are hundreds of distinct rock structures. Geologists find it convenient to divide them into 'primary' and 'secondary' structures.

- Primary Structures: structures formed before or at the same as material is in the process of becoming rock. For example, formed as magma crystallizes or as sediment accumulates.
- Secondary Structures: structures imposed on rock after it has already formed. For example, formed as a result of compression of existing rock.

We'll examine a few of each to see what they tell us.

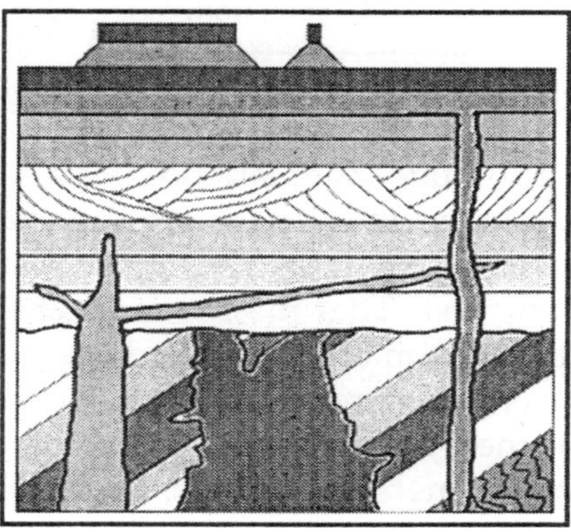

ABOUT THE EARTH

The Earth
As seen from
Space

The third planet from the Sun is your home. The Earth is the only known planet where life can survive. As far as we know, there is no other planet in the universe like Earth. We have a very narrow temperature range that allows water to remain a liquid. Life has developed over millions of years because of that liquid. What else makes us special? We have an atmosphere made up of nitrogen, a relatively inert gas. If we had clouds of sulfuric acid or methane, life may never have developed.

The Earth
As seen from
Space

THE PLATES

There are also huge landmasses on our planet. The rock plates that float across the surface are called tectonic plates. Those plates float on the mantle. The mantle is an area between the core and the crust. That mantle is basically filled with molten rock. It is kept in a liquid state because of the energy given off by the centre (core) of the Earth. Scientists have also discovered that the pressure increases as you move towards the centre of the planet. The core of the Earth has extreme temperatures and pressures that keep the iron and metals liquid and flowing.

The Aurora Borealis
As seen from the
Iss Orbiting Earth

LIQUID IN THE CORE

The flowing metal in our planet helps create something called a dynamo effect. Dynamos create large magnetic fields. In the case of the Earth, the magnetic field protects our planet from space. The region of magnetic protection is called the magnetosphere and protects us from thesolar winds and solar radiation. You can see where solar winds and the magnetosphere collide in the aurora borealis.

TYPES OF ROCKS

Rocks and Minerals

Let's start off with an explanation that rocks and minerals are different things. Rocks are groups of different minerals pushed together and combined. They don't have easy chemical formulas to describe their makeup. A ruby is considered a mineral. It's a nice pretty crystal with the same compounds throughout the object. But the rock that surrounds that ruby has many many different compounds (and even a few pieces of ruby mixed in).

There's an easier example that many people can relate to. Think about quartz. On its own (as a mineral) it's a nice pretty crystal. But a piece of granite often has loads of quartz. It's ground up and crushed, but its still bits of quartz. Granite is a rock and quartz is a mineral.

Rocks Created from Lava and Magma are Defined as Igneous

Igneous Rocks

Igneous rocks are the ones that weresuperheated and originally liquid. They come from the centre of the Earth! Not really the centre, but they often start their lives below the crust and then get pumped out. There are two basic types of igneous rocks. There are the rocks that make it to the surface (extrusive) and the ones that are stuck in the crust just below the surface (intrusive). These igneous types have all hardened after being molten rock. If you walk around a volcano, you will find those extrusive types. The intrusivetypes are usually found in areas called plutons and dikes, big old pools of molten rock that were just beneath the surface. Some examples of igneous rock are granite, all volcanic rock, basalt, and obsidian.

Metamorphic Rocks

This rock type is created by heat and/or pressure. Even though heat is involved, they didn't start off as molten rock. But. You often find metamorphic rock near volcanoes and sources of super hot rock. The heat from the magma changes all of the rock around it. Try another explanation. Look at the name 'metamorphic.' It looks like the word used to describe insects that go through a metamorphosis.

It's the same concept. Some force (heat/pressure) has changed these rocks from one type into a new type. The result is a metamorphic rock. Some examples are marble, jade, slate, and gneiss. Because pressure and heat are involved, these rock types are usually found deep beneath the surface. They are also found near fault lines where plates push against each other and create enormous pressures. Over time, because of the movement of the crust, these metamorphic rocks are pushed to the surface where you can find them every day.

Sedimentary Rocks

The last of the big three rock types is probably the rarest... unless you live near the coast. Sedimentary rock types are created when sediment compresses. It's pretty simple. Here's the setup... A river flows through a canyon and picks up a bunch of silt. That sediment and silt runs downstream and deposits where the river ends. It could be in a flood plain or a valley, but we're using a coastline as an example. When that material gets to the beach, it sits there. Now if you watch this happen over millions of years, more and more sediment builds up and compacts. That compacted sediment eventually becomes a type of rock. Examples of sedimentary rock include sandstone, amber, anthracite, and limestone.

Meteorites

That's right, not all rocks are from Earth. Sure you can debate whether a meteorite is a rock, but you definitely find them all over the Earth. Some super big ones have created massive craters across the globe. Most scientists now believe that a super-massive meteorite hit the Earth and created an extinction event for the dinosaurs. There are also millions of tiny meteorites. They look just like rocks until you break them open. Meteorites are mainly made of iron and nickel. You will also find many smaller trace elements. The key thing to remember about meteorites is that they have much different per centages of elements than rocks on Earth. Many trace elements found on these space travellers are nearly impossible to find on Earth.

TEXTURE AND CLASSIFICATION OF ROCKS

CLASSIFICATION OF ROCKS

A rock is a naturally occurring aggregate of minerals, and certain non-mineral materials such as fossils and glass. Just as minerals are the building blocks of rocks, rocks in turn are the natural building blocks of the Earth's LITHOSPHERE (crust and mantle down to a depth of about 100 km), ASTHENOSPHERE (although this layer, in the depth range from about 100 to 250 km, is partially molten), MESOSPHERE (mantle in the depth range from about 250 to 2900 km), and even part of the CORE (while the outer core is molten, the inner core is solid).

Most rocks now exposed at the surface of the Earth formed in or on continental or oceanic crust. Many such rocks, formed beneath the surface and now exposed at the surface, were delivered to the surface from great depths in the crust and in rare cases from the underlying mantle.

There are two general ways that rocks come to be exposed at the surface:

- Formation at the surface (e.g., crystallization of lava, precipitation of calcite or dolomite from sea water)
- Formation below the surface, followed by tectonic uplift and removal of the overlying material by erosion

There are three major classes of rocks, IGNEOUS, SEDIMENTARY, and METAMORPHIC, with the following attributes:

IGNEOUS ROCKS form by crystallization from molten or partially material, called MAGMA. Magma comes mainly from two places where it is formed,

- In the asthenosphere
- In the base of the crust above subducting lithosphere at a convergent plate boundary.

There are two subclasses of igneous rock, VOLCANIC (sometime called EXTRUSIVE), and PLUTONIC (sometimes called INTRUSIVE).

VOLCANIC ROCKS form at the Earth's surface. They cool and crystallized from magma which has spilled out onto the surface at a volcano. At the surface, the magma is more familiarly known as LAVA.

PLUTONIC ROCKS form from magma that cools and crystallizes beneath the Earth's surface. In a sense, this is the portion of the magma that never makes it to the surface. For the plutonic rock to become exposed at the surface, it must be tectonically uplifted and the overlying material must be removed by erosion.

SEDIMENTARY ROCKS form from material that has accumulated on the Earth's surface. The general term for the process of accumulation is DEPOSITION. The material consists of the products of weathering and erosion, and other materials available at the surface of the Earth, such as organic material. The process by which this otherwise unconsolidated material becomes solidified into rock is variously referred to LITHIFICATION (literally turned into rock), DIAGENESIS or CEMENTATION. Like volcanic rocks, some sedimentary rocks are "lithified" right at the surface, for instance by direct precipitation from sea water. Other sedimentary rocks, like plutonic igneous rocks, are "lithified" below the surface, when they are buried under the weight of overlying sediment. And like the plutonic rocks, sedimentary rocks which were lithified below the surface only become exposed at the surface by tectonic uplift and erosion of the overlying material.

METAMORPHIC ROCKS form when a sedimentary or igneous rock is exposed to high pressure, high temperature, or both, deep below the surface of the Earth. The process, METAMORPHISM, produces fundamental changes in the mineralogy and texture of the rock. The original rock, prior to metamorphism, is referred to as the PROTOLITH.

The protolith can be either an igneous rock or a sedimentary rock, as just indicated. The protolith could also be a previously metamorphosed rock. Ultimately however, if you go far enough back into the history of a metamorphic rock you would find that the first protolith was either a sedimentary or igneous rock. Because all metamorphic rocks form below the surface, for them to become exposed at the surface, they must undergo tectonic uplift and removal of the overlying material by erosion.

In this exercise you will examine some common igneous, sedimentary and metamorphic rocks, and learn to classify them.

CLASSIFICATION

The classification of rocks is based on two criteria, TEXTURE and COMPOSITION. The texture has to do with the sizes and shapes of mineral grains and other constituents in a rock, and how these sizes and shapes relate to each other. Such factors are controlled by the process which formed the rock.

Because igneous, sedimentary, and metamorphic processes are distinct, so too the resulting textures are distinct. Thus there are distinct igneous textures, distinct sedimentary texture, and distinct metamorphic textures. For the purposes of this exercise and routine classification, the kinds of minerals and their proportions, or MINERALOGY, are taken as the natural expression of composition. Fortunately for you, just as the three classes of rocks each have distinct textures, so too do they have distinct mineralogies. Details of TEXTURE and COMPOSITION are discussed in the individual sections on igneous, sedimentary and metamorphic rocks.

Just a note here with regard to grains size. The terms APHANITIC and PHANERITIC mean fine-grained and coarse-grained respectively. Generally, aphanitic means that the grains are too small to see or identify, while phaneritic means that the grains are big enough to see and identify, but the terms are used differently in each the classes of rocks. In igneous rocks the division between aphanitic and phaneritic is taken to be at a grain size of 1/16 mm.

If the grain size is larger than 1/16 mm, the texture is said to be phaneritic. If the grain size is less than 1/16 mm, the texture is said to be aphanitic. In sedimentary rocks, the formal division between aphanitic and phaneritic is taken to be 1/256 mm. For metamorphic rocks the distinction between aphanitic and phaneritic is less quantifiable, but the general meanings are the same.

CLASSIFICATION OF METAMORPHIC ROCKS

Textures

Textures of metamorphic rocks fall into two broad groups,

FOLIATED and NON-FOLIATED. Foliation is produced in a rock by the parallel alignment of platy minerals (e.g., muscovite, biotite, chlorite), needle-like minerals (e.g., hornblende), or tabular minerals (e.g., feldspars). This parallel alignment causes the rock to split easily into thin layers or sheets. Foliation is common in aphanitic as well as phaneritic metamorphic rocks.

Some foliated rocks are also banded. Banding means that the rock consists of alternating, thin layers (typically 1 mm to 1 cm) of two different mineral compositions. Normally, the two types of layers have the same kinds of minerals, but in different proportions, giving the rock a striped appearance. Banding, by itself, defines a foliation.

In order of increasing grain size, foliated textures are referred to as SLATY (aphanitic, very fine grained), PHYLLITIC (aphanitic,fine-grained), SCHISTOSE (phaneritic). The corresponding rock types are called SLATE, PHYLLITE, and SCHIST. These rocks are not normally banded. The composition of the rock (as expressed by it's minerals) is uniform throughout the volume of the rock. The banded, foliated texture is referred to as GNEISSOSE. The corresponding rock type isGNEISS. Gneiss is normally phaneritic, but in some cases the layers are aphanitic.

As the term implies, NON-FOLIATED rocks lack foliation or banding. Such rocks are most commonly composed of minerals that are neither platy nor needle-like, but rather more equidimensional (more or less the same dimension in all directions). Quartz, calcite and dolomite are the most common such minerals. In phaneritic rocks the texture is referred to asGRANOBLASTIC. Common granoblastic rocks included QUARTZITE (quartz), and MARBLE (calcite or dolomite).

The textures of some metamorphic rocks do not fit neatly into any of these categories. In such cases, where the protolith is obvious because the texture of the protolith is well preserved, the prefix "META" is simply used in front of the protolith name, e.g., METAGRANITE, METABASALT, METAWACKE, METACONGLOMERATE, etc.

MINERALOGY

Most of the minerals in igneous rocks and many minerals in sedimentary rock can occur in metamorphic rocks. However, depending on the protolith, many minerals form only during the course of metamorphism. In this sense, minerals such as chlorite, garnet, epidote, staurolite, kyanite, sillimanite, and several others are generally thought of as "metamorphic." A shale protolith (initially clay minerals with lesser amounts of quartz and feldspar) undergoes a number of mineralogical changes when exposed to progressively higher pressures and temperatures during the course of metamorphism. These mineralogical changes are marked by the appearance and growth of metamorphic index minerals at the expense of original minerals (such as clay in shale) and previously former metamorphic minerals.

Starting with shale, the order of appearance of index minerals with increasing grade of metamorphism (from slate to phyllite to schist to gneiss) is typically chlorite, biotite, garnet, staurolite $(Fe,Mg)_2Al_9Si_4O_{22}(O,OH)_2$, kyanite (high pressure Al_2SiO_5) or andalusite (low pressure Al_2SiO_5), sillimanite (high temperature Al_2SiO_5), K-feldspar. Normally, by the time kyanite or andalusite appears, previously formed metamorphic chlorite has long since become unstable and replaced by higher grade minerals. Thus, the kinds of metamorphic minerals in a rock reflect the changing conditions of pressure and temperature that the rock experienced. This kind of information is important for understanding aspects of plate tectonics and the formation of mountain ranges.

TEXTURE OF AN IGNEOUS ROCK

The texture of an igneous rock means the size, shape, and mode of aggregation of its constituent mineral particles. Texture is a very important means of determining the circumstances under which the rock was formed, and hence great attention is paid to it. Since texture responds so accurately to the circumstances of solidification, rate of cooling, pressure, etc., all the varieties shade into one another by imperceptible gradations and form a continuous series.

Nevertheless, it is necessary to distinguish and name the more important kinds. Among the igneous rocks are found four principal types of texture, with several minor varieties:

Glassy

Here the rock is a glass or slag, without distinct minerals in it, though the incipient stages of crystallization, in the form of globules and hair-like rods, are often observable with the microscope. When the glass or slag is made frothy by the bubbles of escaping steam and gas, the texture is said to be vesicular, scoriaceous, or pumiceous (see Figs.), according to the abundance of the bubbles. These are varieties of the glassy texture, though other kinds may also be vesicular. A vesicular rock in which the steam-holes have been filled up by the subsequent deposition of some mineral is called amygdaloidal, a term derived from the Greek word for almond.

The Compact (Or Felsitic)

The Compact (Or Felsitic) texture is characterized by the formation of exceedingly minute crystals, too small to be seen by the unassisted eye, giving the rock a homogeneous but stony and not glassy appearance. If the crystals are too minute to be identified even by the aid of the microscope, the rock is said to be cryptocrystalline, and when such identification can be made, it is called micro crystalline.

Porphyritic

In rocks of this texture are large, isolated crystals, called phenocrysts, embedded in a ground mass, which may be glassy or made up of small crystals. The phenocrysts may have sharp edges and well-formed faces, or they may have irregular and corroded surfaces. The porphyritic texture indicates two distinct phases of crystallization. The first is the formation of the phenocrysts, which remain suspended in the molten mass, or magma, and are often corroded and partially redissolved (resorbed) by it.

These crystals are said to be of intratelluric origin, because formed before the eruption of the lava, and such crystals are showered out of certain active volcanoes at the present time. Stromboli for example, ejects quantities of large and perfect augite crystals. There is reason to believe, however, that not all phenocrysts are thus intratelluric, but that the first phase of crystallization sometimes takes place after the ejection of the molten mass. The second phase consists in the formation of the ground mass, which may be glassy, finely crystalline, or both. Mineral particles having distinct crystalline form are called idiomorphic.

Granitoid

In this texture the rock is wholly crystalline, without ground mass or interstitial paste. The component grains, which may be fine or very coarse, are of quite uniform size, and as the crystals have interfered with one another in the process of formation, they have rarely acquired their proper crystalline shape. Such grains are said to be allotriomorphic.

An additional texture which should be mentioned is the frag-mental. This is represented by the accumulations of the frag-mental products ejected by volcanoes, agglomerates, bombs, lapilli, ashes, etc. Many such materials accumulate in bodies of water and are there sorted and stratified and, it may be, mingled with more or less sand and mud and other sedimentary material. Rocks formed in this manner partake of the nature of both the igneous and sedimentary classes, and may be regarded as a series intermediate between the other two and in a measure connecting them. These rocks will here be treated as a special subdivision, under the name of pyroclastic rocks.

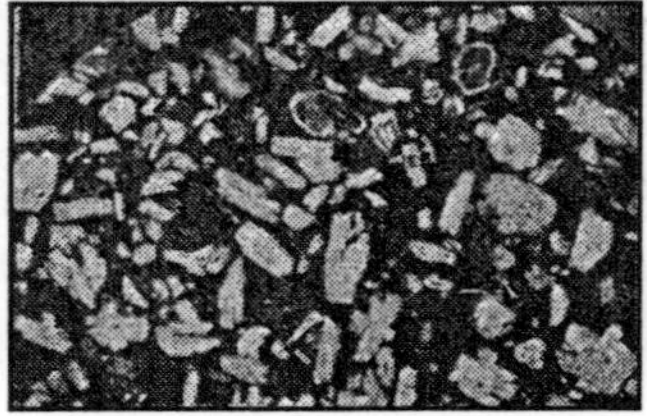

Fig. Slab of Polished Porphyry, Natural Size Phenocrysts of Felspar.

In our studies of the products of modern volcanoes, we

saw that the same molten mass will give rise to rocks of very different appearance in its different parts, according to the circumstances of rapidity of cooling, pressure, etc. We may now express this in somewhat more general form and say that the texture of an igneous rock is determined by the several factors which affect the molten mass during consolidation. Of such factors may be mentioned the chemical composition, temperature, rate of cooling, degree of pressure, and the quantity present of dissolved vapours and gases, which are called mineralizers.

Fig. Hand Specimen of Granite, Natural size.

The names and terminology applied to igneous rocks can be quite confusing for geology students. Various terms may refer to rock texture, mineral constituents, or chemical composition. Many names with vague or poorly defined meanings have been applied over the years to the great variety of rocks formed by cooling from magma or lava. Some of these names are quite strange sounding—trachyte, while others are in the common English vocabulary—granite.

From this often confusing situation, a commonly used classification for igneous rocks has emerged based on two criteria—texture and chemical composition. Texture refers to the size of crystals, presence of glass, and porosity of the rock. Texture is determined primarily by how the magma or lava cooled. Following are common textural terms for igneous rocks.

Phaneritic: Large crystals that are clearly visible to the eye with or without a 10-power hand lens. The entire rock is made up of large crystals, which are generally ½ mm to cm in size;

no fine matrix material is present. This texture forms by slow cooling of magma deep underground in the plutonic environment.

Aphanitic: Small crystals that cannot be seen by the eye with or without a 10-power hand lens. The entire rock is made up of small crystals, which are generally less than ½ mm in size. This texture results from rapid cooling involcanic or hypabyssal (shallow subsurface) environments.

Porphyritic: Texture in which an aphanitic matrix makes up part of the rock, and large crystals are present within the matrix. The large crystals are called phenocrysts. This texture implies two-stage cooling—an early stage of slow cooling in which the phenocrysts grow, followed by a later stage of rapid cooling that forms the matrix.

Pegmatitic: Pegmatites are composed of quite large crystals—cm to tens of cm in size. They typically occupy veins or layers within a larger plutonic body. The large crystals form by slow cooling of magma.

Glassy: Non-crystalline (glassy) structure of the rock, in which no minerals are present. Glass results from cooling that is so fast that minerals do not have a chance to crystallize. This may happen when magma or lava comes into quick contact with much cooler materials near the Earth's surface. Pure volcanic glass is known as obsidian.

Vesicular: This term refers to vesicles (holes, pores, or cavities) within the igneous rock. Vesicles are the result of gas expansion (bubbles), which often occurs during volcanic eruptions. Pumice and scoria are common types of vesicular rocks.

Breccia: A rock composed of broken, angular fragments of mixed composition. Such texture forms in volcanoes, along fault zones, and in landslides.

Igneous Rock Composition

Classification based on chemistry takes into account the amount of silica (SiO_2) and the composition of feldspar minerals (K, Na, Ca). Igneous rock chemistry is determined mainly by the source of the magma and any interactions between magma and the rocks through which it migrates. Chemical composition usually is indicated by the minerals or colour of an igneous rock. Four main compositional categories result from this approach.

Felsic: Rich in feldspars and silica. Silica content ranges from about 55% to > 70%. Potassium feldspar makes up more than one-third of total feldspars; plagioclase (Na and Ca) feldspars are less than two-thirds of total feldspars. Typical of continental crust.

Intermediate: Between felsic and mafic. Silica content ranges from about 55% to 65%. Plagioclase feldspars make up more than two-thirds of total feldspars. Na-rich plagioclase predominates over Ca-rich plagioclase. Found in association with subduction zones.

Mafic: Rich in magnesium and iron with less silica. Silica content is 45% to 50%. Ca-rich plagioclase is the dominant feldspar with little or no K- or Na-feldspars. Typical of oceanic crust.

Ultramafic: Still more magnesium and iron and even less silica. Silica content is less than 45%, and little or no feldspar is present. Derived from the mantle.

IGNEOUS ROCK CLASSIFICATION

Texture combined with chemical composition is the basis for modern classification of igneous rocks. Although chemical composition cannot be "seen" in the field, it can be estimated based on the main minerals of a rock specimen. The classification below considers the amount of quartz and the types of feldspar (K, Na, Ca) varieties as primary indicators for chemical composition.

Table.Classification of Common Igneous Rocks.

Composition	Phaneritic (main minerals)	Aphanitic	Color
Felsic	Granite (>10% quartz, >2/3s K-feldspar)	Rhyolite	10
	Syenite (<10% quartz, >2/3s K-feldspar)	Trachyte	15
	Monzonite (1/3 to 2/3s K-feldspar)	Latite	20
Intermediate	Granodiorite (>10% quartz, >10% K-spar, >2/3s Na-spar)	Dacite	20
	Diorite (<10% quartz, <10% K-feldspar, >2/3s Na-feldspar)	Andesite	25
Mafic	Gabbro (<10% quartz, >2/3s Ca-feldspar, olivine)	Basalt Lamprophyre	50
Ultramafic	Peridotite (pyroxene and olivine; no quartz or feldspar)		95

Lamprophyre is a special igneous rock of mafic composition. It is porphyritic and occurs in shallow (hypabyssal) intrusions. The phenocrysts consist of mafic minerals (biotite, hornblende and pyroxene); the same mafic minerals are found in the fine groundmass along with feldspars. Some dikes of the Spanish Peaks region are lamprophyres.

The colour index (right column of table) indicates the per centage of dark-coloured minerals in the rock. Dark-coloured minerals include olivine, pyroxene, biotite, hornblende, and iron oxides, which are generally dark green to black. Quartz, feldspars, and muscovite are light-coloured. This includes pink and red potassium feldspars as well as gray Ca-rich plagioclase. Colour index is somewhat variable and works best with phaneritic rocks. The colour index is more difficult to apply to aphanitic rocks, as their hue and brightness vary considerably. For example, rhyolite may be pale gray, green, or pink to dark red. Andesite is usually moderate gray or green, and basalt is dark gray to black. However, the colours for glassy rocks (obsidian, scoria, pumice) are meaningless as indicators for chemical composition.

The classification of igneous rocks has been the subject of frequent debate and voluminous literatuie. Over the past decade, most geologists have accepted the IUGS (International Union of the Geological Sciences) classification as the standard. Since this classification is being widely adopted, it bears discussion. However, as we shall see is rather complex and best left to advanced students. For our purposes, we will introduce and discuss a much simpler classification that will allow us to easily identify the more common igneous rocks.

Classifying Things

Carolus Linneaus proposed the first classifcation for biological organisms in the 18th century. This taxonomic classification was designed to simplify the complexity of nature by lumping together living species that shared common traits. So to classifications in the earth sciences are designed to reduce complexity. For instance, the classification of minerals is based on common anoins since minerals sharing common anions often have similar physical properties (i.e hardness, cleavage etc.). Rock classifications also seek to reduce complexity. Most are what we term genetic. That means that by pigeonholing a rock in a certain group we say something about its genesis or origin. For example, aphanitic rocks are or volcanic origin while phaneritic rocks are plutonic.

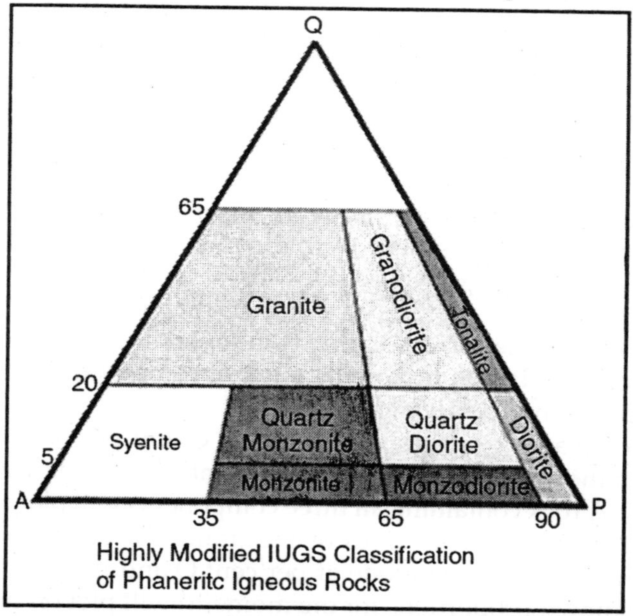

Highly Modified IUGS Classification of Phaneritc Igneous Rocks

Igneous rocks are classified on the basis of mineralogy, chemistry, and texture. As discussed earlier, texture is used to subdivide igneous rocks into two major groups: the plutonic rocks, with mineral grain sizes that are visible to the naked eye, and the volcanic rocks, which are usually too fine-grained or glassy for their mineral composition to be observed without the use of a petrographic microscope. As noted in the sidebar to the left, this is largely a genetic classification based on the depth of origin of the rock (volcanic at or near the surface, and plutonic at depth). Remember that porphyritic rocks have spent time in both worlds. Let's first examine the classification of plutonic rocks.

A plutonic rock may be classified mineralogically based on the actual proportion of the various minerals of which it is composed (called the mode). In any classification scheme, boundaries between classes are set arbitrarily. The International Union of Geological Sciences (IUGS) Subcommission on the Systematics of Igneous Rocks in 1973 suggested the use of the modal composition for all plutonic igneous rocks with a colour index less than (Image to the right). A second scheme (not shown) was proposed for those plutonic ultramafic rocks with a colour index greater than.

The plotting of rock modes on these triangular diagrams is simpler than it may appear. The three components, Q (quartz) + A (alkali (Na-K) feldspar) + P (plagioclase), are recalculated from the mode to sum to per cent. Each component is represented by the corners of the equilateral triangle, the length of whose sides are divided into equal parts. Any composition plotting at a corner, therefore, has a mode of per cent of the corresponding component. Any point on the sides of the triangle represents a mode composed of the two adjacent corner components.

For example, a rock with per cent Q and per cent A will plot on the QA side at a location per cent of the distance from A to Q. A rock containing all three components will plot within the triangle. Since the sides of the triangle are divided into parts, a rock having a mode of per cent Q and per cent A + P (in unknown proportions for the moment) will plot on the line that parallels the AP side and lies per cent of the distance toward Q from the side AP. If this same rock has per cent P and per cent A, the rock mode will plot at the intersection of the per cent Q line described above, with a line paralleling the QA side at a distance per cent toward P from the QA side. The third intersecting line for the point is necessarily the line paralleling the QP side at per cent of the distance from the side QP toward A. A rock with per cent Q, per cent P, and per cent A plots in the granite field, whereas one with per cent Q, per cent P, and per cent A plots in the granodiorite field. The latter is close to the average composition of the continental crust of the Earth.

Ideally it would be preferable to use the same modal scheme for volcanic rocks. However, owing to the aphanitic texture of volcanic rocks, their modes cannot be readily determined; consequently, a chemical classification is widely accepted and employed by most petrologists. One popular scheme is based on the use of both chemical components and normative mineralogy. Because most lay people have little access to analytic facilities that yield igneous rock compositions, only an outline will be presented here in order to provide an appreciation for the classification scheme.

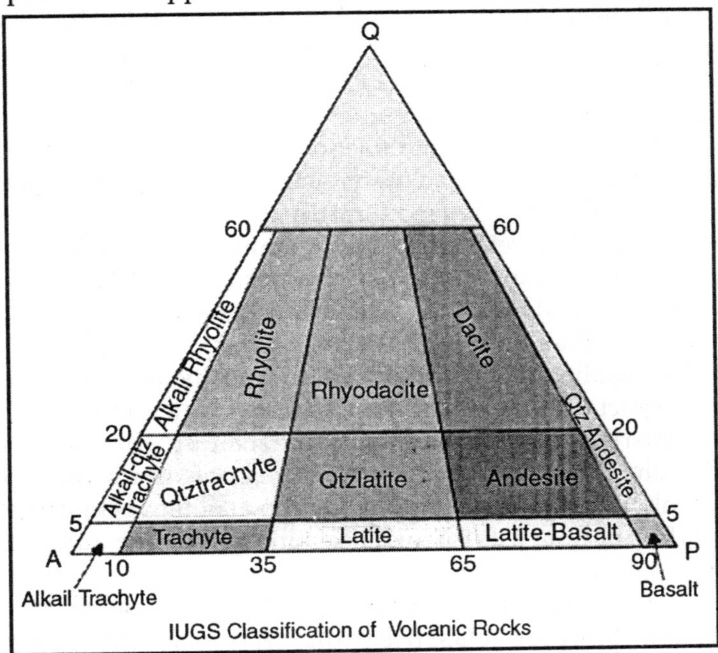

IUGS Classification of Volcanic Rocks

The major division of volcanic rocks is based on the alkali (soda + potash) and silica contents, which yield two groups, the subalkaline and alkaline rocks. Furthermore as they are so common, the subalkaline rocks have two divisions based mainly on the iron content with the iron-rich group called the tholeiitic series and the iron-poor group called calc-alkaline. The former group is most commonly found along the oceanic ridges and on the ocean floor and is usually restricted to mafic igneous rocks like basalt and gabbro; the latter group is characteristic of the volcanic regions of the continental margins (convergent, or destructive, plate boundaries) and is comprised of a much more diverse suite of rocks.

Chemically the subalkaline rocks are saturated with respect to silica. This chemical property is reflected in the mode of the mafic members that have two pyroxenes, hypersthene and augite [Ca(Mg, Fe)Si2O6], and perhaps quartz. Plagioclase is common in phenocrysts, but it can also occur in the matrix along with the pyroxenes. In addition to the differences in iron content between the tholeiitic and calc-alkaline series, the latter has a higher alumina content (16 to 20 per cent), and the range in silica content is larger (48 to 75 per cent compared to 45 to 63 per cent for the former).

Hornblende and biotite phenocrysts are common in calc-alkaline andesites and dacites but are lacking in the tholeiites. Dacites and rhyolites commonly have phenocrysts of plagioclase, alkali feldspar (usually sanidine), and quartz in a glassy matrix. Hornblende and plagioclase phenocrysts are more widespread in dacites than in rhyolites, which have more biotite and alkali feldspar.

The alkaline rocks typically are chemically undersaturated with respect to silica; hence, they have only one pyroxene, the calcium-rich augite) and lack quartz but often have a feldspathoid mineral, nepheline. Microscopic examination of alkali olivine basalts (the most common alkaline rock) usually reveals phenocrysts of olivine, one pyroxene (augite), plagioclase and perhaps nepheline.

Now that we have completely confused you,let's look at a much simpler classification. We call this a *field classification* because it requires little detailed knowledge of rocks and can be easily applied to any igneous rock we might pick up while on a field trip. It utilizes texture, mineralogy and colour. The latter is a particularly unreliable property, but the classification realises that certain fine-grained (aphanitic) igneous rocks contain no visible mineral grains and in their absence colour is the only other available property. Students the thus cautioned to use colour only as a last resort.

Texture	Felsic (Light Color)	Intermediate	Mafic (Darl color)	Ultamafic
Coarse	Granite	Diorite	Gabbro	Peridotite
Fine	Rhyolite	Andesite	Basalt	
Vesi-Cular	Pumice		Scoria	
Glassy	Obsidian			
	Minerals Present			
	Quartz K-Feldspar NA-Plag	NA-CA-Plag Amphibole	CA Plag Pyroxene	Pyroxene Olivine

To employ this classification we must first determine the rock's texture. if you wish to go to a discussion of igneous rock textures. You might remember we have five basic textures; phaneritic (coarse), aphanitic (fine), vesicular, glassy and fragmental (our classification doesn't bother with the latter because we often term all fragmental igneous rocks tuffs). Examine your rock and determine which textural group it belows to.

If it is glassy, vesicular or fragmental you cannot determine mineralogy and hence the name is simply obsidian for a glass, tuff for a fragmental or pumice/scoria for a vesicular rock (the latter are differentiated on the basis or colour and size of the vesicles or holes).

For the phaneritic and some aphanitic rocks you must determine the mineralogy. Often it is only necessary to identify one or two key minerals, not all of the minerals in the rock. For instance quartz and potassium feldspar (k-feldspar) are restricted to granites and rhyolites.

Amphibole is only abundant in diorite or andesite, although minor amounts can be present in granite. How am I getting these names?

Let's take an example. I pick up my first specimen and notice that it is distinctly coarse grained (phaneritic). This means that it must be one of the rocks in the row labeled *coarse* (i.e., granite, diorite, gabbro or peridotite). I next place the rock under a binocular microscope and identify the minerals plagioclase and pyroxene.

I go to the bottom row of the chart (Minerals Present) and look for a match with my mineralogy. I find it in the third column (Ca-play, pyroxene) and read the name (gabbro) from the *coarse* row on the chart. Pretty simple!! Relax, when you actually begin your igneous rock identification we will walk you through it step by step. But remember to refer to the above classification diagram often as an aid.

Chapter 3

Stratification

INTRODUCTION

STRATIFICATION AND SEGREGATION

Stratification describes the way in which different groups of people are placed within society. The status of people is often determined by how society is stratified - the basis of which can include;

Stratification - strata - meaning layers geological stratification - layers in geological terms, ie rock layers best seen in cliff faces, where you can view different bands or layers of rock, sand, stone etc. also shows an anticline or syncline (layers tipping upwards or downwards) showing geological forces at work.

Can also refer to how deposits are formed that created the strata (layers) colin and gus have explained it very well:

- *Wealth and income*: This is the most common basis of stratification
- Social class
- Ethnicity
- Gender
- Political status
- Religion (e.g. the caste system in India)

The stratification of society is also based upon either an open, or closed, system.

Open

Status is achieved through merit, and effort. This is

sometimes known as a meritocracy. The UK is a relatively open society, although disadvantaged groups within society face a glass ceiling.

Closed

Status is ascribed, rather than achieved. Ascribed status can be based upon several factors, such as family background (e.g. the feudal system consists of landowners and serfs). Political factors may also play a role (e.g. societies organised on the basis of communism), as can ethnicity (e.g. the former apartheid regime in South Africa) and religion.

A technique used to analyse/divide a universe of data into homogeneous groups (strata) often data collected about a problem or event represents multiple sources that need to treated separately. It involves looking at process data, splitting it into distinct layers (almost like rock is stratified) and doing analysis to possibly see a different process.

For instance, you may process loans at your company. Once you stratify by loan size (e.g. less than 10 million, greater than 10 million), you may see that the central tendency metrics are completely different which would indicate that you have two entirely different processes...maybe only one of the processes is broken.

Stratification is related to, but different from, Segmentation.

A stratifying factor, also referred to as stratification or a stratifier, is a factor that can be used to separate data into subgroups. This is done to investigate whether that factor is a significant special cause factor.

PAUCITY OF STRATIFICATION

Being deposited in avalanche fashion, the pumice flows usually show no trace of stratification and at best a crude one. Here and there thin lenses and stringers of fine tuff among the coarser materials suggest the settling of swirls of dust during lulls in the passage of successive flows.

Locally, also, there is a rough alignment of the large lumps near the base of the flows, probably caused by deformation under load. Apparently, however, the pumice lumps were neither hot enough nor sufficiently rich in gas to suffer welding.

Bedding is most conspicuous in the canyons of Annie, Sun, and Sand creeks, where the earlier deposits of pale dacite pumice are overlain by a darker layer of basic scoria. Within individual flows there is usually no stratification whatever. See figure.

Description

Stratification is a technique used in combination with other data analysis tools. When data from a variety of sources or categories have been lumped together, the meaning of the data can be impossible to see. This technique separates the data so that patterns can be seen.

USING STRATIFICATION

Before Collecting Data.

When data come from several sources or conditions, such as shifts, days of the week, suppliers or population groups.

When data analysis may require separating different sources or conditions.

STRATIFICATION PROCEDURE

Before collecting data, consider which information about the sources of the data might have an effect on the results. Set up the data collection so that you collect that information as well.

When plotting or graphing the collected data on a scatter diagram, control chart, histogram or other analysis tool, use different marks or colours to distinguish data from various sources. Data that are distinguished in this way are said to be "stratified."

Analyse the subsets of stratified data separately. For example, on a scatter diagram where data are stratified into data from source and data from source, draw quadrants, count points and determine the critical value only for the data from source, then only for the data from source.

STRATIFICATION EXAMPLE

The ZZ-400 manufacturing team drew a scatter diagram to test whether product purity and iron contamination were related, but the plot did not demonstrate a relationship. Then a team member realized that the data came from three different reactors.

The team member redrew the diagram, using a different symbol for each reactor's data:

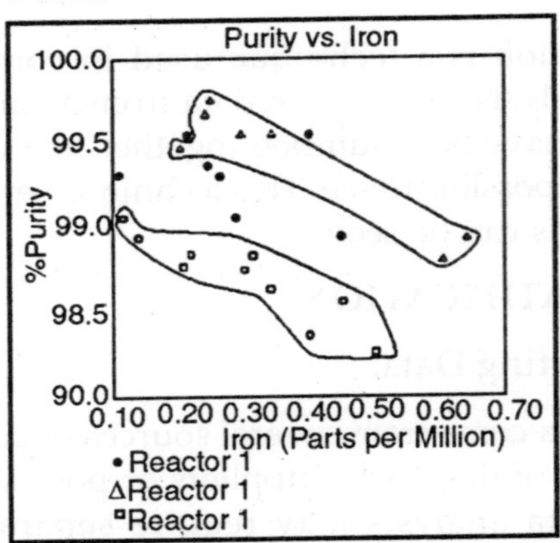

Now patterns can be seen. The data from reactor and reactor are circled. Even without doing any calculations, it is clear that for those two reactors, purity decreases as iron increases. However, the data from reactor, the solid dots that are not circled, do not show that relationship. Something is different about reactor.

STRATIFICATION CONSIDERATIONS

Here are examples of different sources that might require data to be stratified:

- Equipment
- Shifts
- Departments
- Materials
- Suppliers
- Day of the week
- Time of day
- Products
- Survey data usually benefit from stratification.

Always consider before collecting data whether stratification might be needed during analysis. Plan to collect stratification information. After the data are collected it might be too late. On your graph or chart, include a legend that identifies the marks or colours used.

CREATE A STRATIFICATION DIAGRAM

Stratification Diagram (Excel, 78 KB) – Analyse data collected from various sources to reveal patterns or relationships often missed by other data analysis techniques. By using unique symbols for each source, you can view data sets independently or in correlation to other data sets.

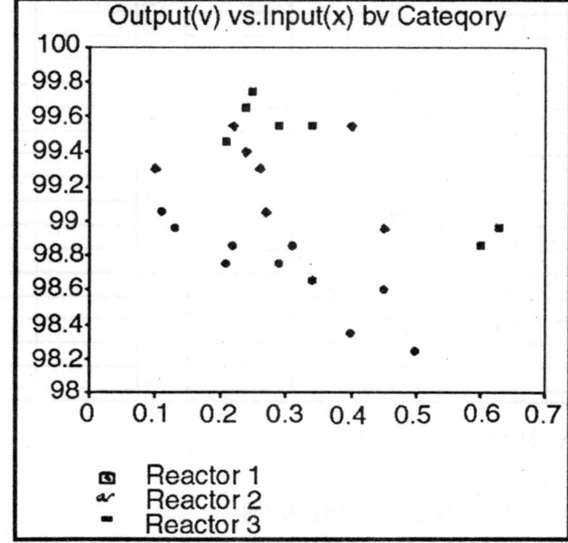

	Reactor 1		Reactor 2		Reactor 3	
	Input (x)	Output (y)	Input (x)	Output (y)	Input (x)	Output (y)
1	0.45	98.95	0.6	98.85	0.5	98.25
2	0.27	99.05	0.63	98.95	0.4	98.35
3	0.26	99.3	0.21	99.45	0.45	98.6
4	0.1	99.3	0.29	99.55	0.34	98.65
5	0.24	99.4	0.34	99.55	0.29	98.75
6	0.4	99.55	0.24	99.65	0.21	98.75
7	0.22	99.55	0.25	99.75	0.31	98.85
8					0.22	98.85
9					0.13	98.95
10					0.11	99.05
11						
12						
13						
14						
15						
16						
17						
18						
19						
20						

	Category 4		Category 5		Category 6	
	Input (x)	Output (y)	Input (x)	Output (y)	Input (x)	Output (y)
1						
2						
3						
4						
5						
6						
7						
8						
9						
10						
11						
12						
13						
14						
15						
16						
17						
18						
19						
20						

LAMINATION BEDDING

EXPERIMENTS ON LAMINATION OF SEDIMENTS

Sedimentology: Experiments on lamination of sediments,

resulting from a periodic graded-bedding subsequent to deposition—a contribution to the explanation of lamination of various sediments and sedimentary rocks.

These sedimentation experiments have been conducted in still water with a continuous supply of heterogranular material.

A deposit is obtained, giving the illusion of successive beds or laminae. These laminae are the result of a spontaneous, periodic and continuous grading process, which takes place immediately, following the deposition of the heterogranular mixture.

The thickness of the laminae appears to be independent of the speed of sedimentation but increases with extreme differences in the size of the particles in the mixture. Where a horizontal current is involved, thin laminated superposed layers developing laterally in the direction of the current, are observed.

Laminae have traditionally been considered by geologists as strata with a thickness of less than 1 cm. Augustin Lombard gives the following definitions: 'Lamination groups together all the structures characterising sedimentary rocks within a bed or stratum. Beds or laminations are the internal arrangement of strata in lithologically distinct levels.' A stratum is a sedimentary unit included between two boundary surfaces.

It consists of a deposit of sediments of various structures accumulated during a continuous phase. Regarding its genesis, Lombard writes: 'The origin of planar-parallel laminae is attributed to currents during deposition.' 'Repetitions of graded-bedding in a stratum seems to result from successive pulsations in the flow of the mass of sediment.' However, Lombard also writes: 'Kuenen has reproduced these laminae without any current pulsations.

They are formed during decelerations.'[5] Having read this, I wondered whether the presence of a current was necessary in the formation of laminae, and whether these laminae could not result just as well from a continuous sedimentation in still water.

FUNDAMENTAL EXPERIMENTS

To test this hypothesis, I performed three very simple experiments:

- A mixture was prepared consisting of 25% sand with particles ranging in size between 0.3 mm and 0.4 mm coloured with methylene blue, and 75% siliceous powder with particles between 20 and 80 microns. Then, during a period of 10 minutes the dry mixture was poured into a 2 litre conical-shaped vessel. shows that an alluvium cone was formed in the vessel, as a result of the mixture having fallen into the vessel at a constant rate.

Almost horizontal graded sequences appeared simultaneously at the bottom of the vessel, composed of blue sand underneath and siliceous powder of approximately 2.5 mm thickness on top.

A striking parallel can be obtained by a dry flow of mixtures of powders.' These experiments show segregation of particles of the same size.

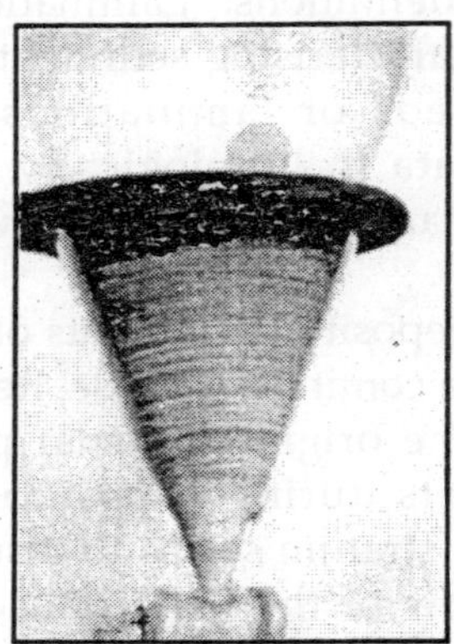

Fig. Laminations Resulting from Flowing of dry Sediment.

- 5 kg of the same mixture was poured into the funnel of a screw powder distributor, which provides a continuous flow of the mixture at varying speeds into a 2 litre test tube full of water. 400 g of the mixture was poured into the test tube at a speed of 140 g per hour. A laminated deposit was obtained. The thickness of the laminae was practically constant at about 2.5 mm. A cross-section through the dry deposit showed the laminae.

- In order to observe the lamination mechanism better, the previous experiment was repeated with the water being coloured with methylene blue. shows that the top part of the resultant deposit is clearer than the rest of the deposit. This top part forms a ring, composed mostly of siliceous powder, into which particles of blue sand can be seen to penetrate. As the ring grows thicker, a thin layer of blue sand appears within it. This constitutes the base of a new lamina. And so it goes on.

The genesis of this lamination, therefore, results quite clearly from a segregation of particles of the same size within the deposited mixture. Graded bedding has occurred within the deposit itself, after its sedimentation. There is no superposition of layers.

EFFECT OF SEDIMENTATION SPEED

Does the speed of sedimentation cause a variation in the thickness of the laminae?

Sand calibrated between 0.135 mm and 0.400 mm coloured with blue methylene, and a siliceous powder between 20 and 40 microns were used.

Fig. Laminations Resulting from Flowing in Water.

Variation in Sedimentation Speed

Successively, 50, 100, 200 and 300 cm^3 of powder were diluted in a test tube full of water. The sedimentation time increased according to the increasing quantity of powder being diluted. Into each of these dilutions, the same 80 cm^3 volume of sand flowed from the distributor during 5 minutes. The thickness of the resulting laminae remained constant at about 2.5 mm.

Variation in the Sedimentation Speed of Sand

40, 80 and 120 cm^3 of sand were poured successively for 5 minutes into the same dilution of 200 cm^3 of siliceous powder in 2 litres of water. The thickness of the resulting strata remained constant at about 2.5 mm.

EFFECT OF PARTICLE SIZE

Do laminae vary in thickness with extreme differences in the size of the particles?

Variation of Thickness with the size of large Particles of Sand

200 cm^3 of siliceous powder from 20 to 80 microns diluted in a test tube of water, had poured into it for a period of 5 minutes on each occasion, 100 cm^3 of sand, with the size of the larger particles increasing each time. The thickness of the resulting laminae increased with the widening difference between the size of the particles of sand in each discharge as follows:

Size range of sand-mm	Thickness of laminations-mm
0.125-0.160	2
0.200-0.250	2
0.315-0.400	2.5
0.500-0.630	2.7
0.630-0.800	3
0.800-1.000	laminations did not form

Variation in Thickness due to Sizes of Siliceous Powder Particles

100 cm^3 of sand with particles between 0.3 mm and 0.4 mm was mixed successively with six portions of 200 cm^3 of siliceous powder and sand, whose particles were increased in size each time.

The mixture was poured for five minutes each into separate flasks and the laminations were observed. The thickness of the laminae increased with the size of the particles of the fine powder, but not to any great extent as follows:

Size range of siliceous powder-microns	Thickness of laminations-mm
0-4	2.5
20-40	2.5
63-80	2.5
80-100	2.5
125-160	2.7
200-250	2.7

Variation in Thickness due to wide Difference in size of Particles of sand

Four measures of sand were poured successively from the distributor. The size of the small particles of sand in each measure remained the same, but the larger particles were increased in size for each measure. shows that the thickness of the laminae increased as the difference between the size of particles became greater as follows:

Size of small sand particles-microns	Size of large sand particles-mm	Thickness of laminations-mm
4	0.4	5
4	0.63	6
4	0.8	8
4	1	10

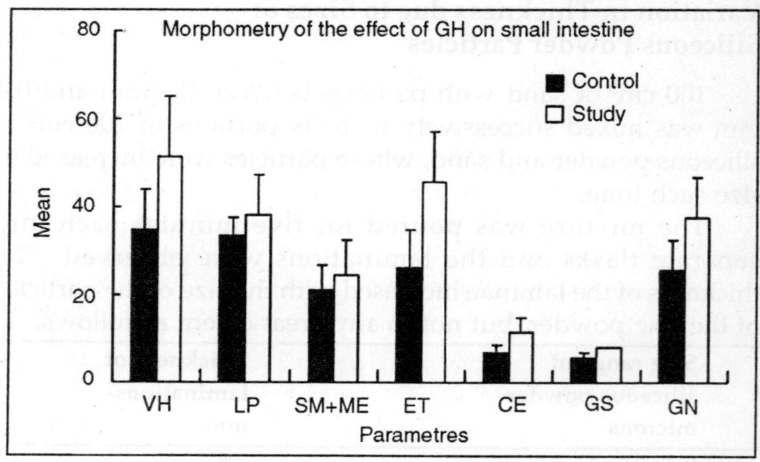

Fig. The Thickness of Laminae is Independent of the Speed of Sedimentation.

Three further experiments with larger differences in size (1.25 mm–1.6 mm–2 mm) gave irregular lamination. The obvious conclusion from this second series of experiments is that the thickness of laminae increases with an increase in the difference between the size of the particles in them.

Simultaneous Variations of Sedimentation

A mixture, composed of 25% sand and 75% siliceous powder, flowed from the distributor into the test tube full of water, at a speed of 140g per hour, then at double and triple this speed. shows that the thickness of the laminae remained constant.

The obvious conclusion is that within experimental limits the thickness of laminae is independent of the sedimentation speed.

NATURAL LAMINATED SEDIMENTARY ROCKS

Numerous fluviatile and marine sediments, as well as sedimentary rocks, showing the microstratified aspect are given such names as: laminae, laminites, varves, etc. These types of lamination are attributed to successive deposits of layers.

The question is whether some of these natural laminae can be explained by the mechanism demonstrated above. In this connection the experimental method was applied to the following natural sedimentary rocks found in France.

- A multicoloured sandstone from Fontainbleau shows lamination of an average of 3 mm thickness. It was crumbled into particles varying from 0.1 to 0.3 mm in size. The particles were placed in the distributor, and fed into the 2 litre test tube full of water at a speed of 50 g per hour. The original lamination was reproduced quite evenly with an almost identical thickness.
- A diatomite from Auvergne shows lamination. Reduced to its elementary particles it was calibrated by a cyclone and particles of at least 80 microns were selected. Examination by microscope confirmed that the majority of the diatoms were unbroken. The larger particles were coloured with methylene blue and then mixed with the smaller particles.

Fig. Sample of Diatomite.

The mixture was fed from the distributor into the test tube of water at three successive speeds of 50, 100 and 150g per hour for identical periods of time. Lamination appeared in the deposit and the thickness did not vary with the sedimentation speed. The original lamination was reproduced with virtually the same thickness.

INCIDENCE OF LATERAL CURRENT

A plate measuring 80 cm by 25 cm by 40 cm was fixed at one end of a flume, just at the water level. It received a horizontal current of water at a slow speed. Coming from the flume through a circulation pump, and at the same time the diatomite particles were falling at a speed of 80 g per hour. The experiment was continuous for 15 days, apart from one interruption.

Thin laminated superposed layers expanding laterally in the direction of the current were observed. They could be distinguished from each other by a gradation of colours, which were clearly due to the different loads of large coloured particles.

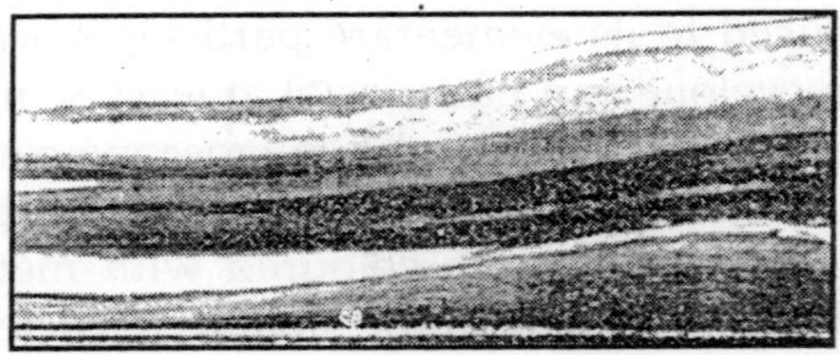

Fig. Lamination Resulting from a Lateral water Flow.

CONCLUSIONS

The continuous deposit of a heterogranular sediment in still water was studied:

- It was noted that the deposited material organised itself immediately after deposition into periodic graded laminae giving the appearance of successive beds.
- One of the more striking features of these laminae formed in the sediment itself was their regular periodicity.
- The thickness of the laminae was measured in millimetres. It was independent of the speed of sedimentation and varied according to the extreme difference in the size of the mixed particles.
- When deposition took place in a water flow, the lamination phenomenon was also observed. The

geometry of lamination was modified by the water flow, but the latter was not the cause of the modification.

- The periodic graded laminae were similar to the laminae or varves observed in nature which are interpreted as a superposition of seasonal or annual beds. Their origin, however, was quite different, arising from periodic structuring after deposition.
- The question now is to study a number of laminated or varved formations in relation to this mechanism, particularly looking for physical structuring obtained from experimentation.

Fig. Laminations Resulting from the Resedimentation of the Diatomite.

The answer to the question is in the affirmative, that is, natural laminae can be explained by this demonstrated mechanism. Notwithstanding the case where it can be demonstrated that lamination results from seasonal deposits.

OUTCROP

LASER OUTCROP MAPPING

Laser outcrop mapping is being recognized as the future of outcrop field geology. Laser mapping utilizes LIDAR (light detection and ranging) and various forms of image capture, laser intensity, traditional photography and digital imaging. The concept is simple. A laser is used to determine the position of points along an outcrop face in x, y, and z space. If a sufficient number of laser points are gathered along an outcrop, the shape of the outcrop will be approximated by the point cloud.

The points can then connected into any type of mesh or 3-D grid to construct a surface in much the same way a seismic horizon is interpolated from interpretation on individual seismic lines. Any image(s) can be applied to the surface as a texture. The result is a 3-D surface model with a series of images fit on to the surface. A "virtual outcrop" is born.

In order to have a high-resolution virtual outcrop, first you need a high-resolution terrain model. The tighter the point cloud is the better and bigger the terrain model will be. Some high-end laser scanners collect not only x, y, and z data but also the intensity of the reflected laser pulse.

The result of laser intensity plus x, y, z data looks like a black and white photo without sun shadows. The intensity "black and white" can be used to rectify and "cookie-cut" another image onto the model, a digital photo for instance that has been blended with laser intensity.

All of these techniques can be done with little or now trouble as long as the data set in question is small enough to fit on your computer. A standard texture-mapped VRML model works well across platforms in most web-browsers for mass consumption.

The "quality" of the model is relevant to the scale of the outcrop under investigation and the purpose for which the data are to be used. If the desired output model needs to be at 1 cm point spacing and the outcrop is 3 km long and 100 m high, the file size of only the x, y, z and intensity data is 42 gigabytes binary! The obvious solution then is to decimate your terrain data.

Our procedure extracts the full resolution data from the intensity, splits off the x, y, z data, optimizes the x, y, z data and then rejoins them together as a single data set. This process can take a 42 gigabyte dataset down to 20 megabytes while preserving great detail of the original outcrop in full navigable 3-D.

One of the ways we have been utilizing the data at the Bureau of Economic Geology at the University of Texas at Austin's Jackson School of Geosciences is to treat the data in much the same manner as we would seismic attribute data. Examine multiple attributes simultaneously and look for patterns mathematically to aid in bringing out the subtleties of the geology that might be overlooked otherwise.

One example is this data set which used a combination of first derivative (slope) of the outcrop as well as laser intensity to perform a multi-component classification. This technique distinguishes between thin sands and shale as well as differences in sand grain mineralogy.

Fig. Laser Point Cloud Data Set at 1 cm xyz Point Spacing. Human in Front near Outcrop is Approximately 6 Feet Tall.

Fig. Full Resolution Laser Intensity Point Cloud. Human for Scale in Foreground Near Outcrop is Approximately 6 Feet Tall.

Fig. Laser Intensity Image Blended with a Digital Photo and Applied as a Texture to the xyz Model.

Fig. This is a Multi-component Analysis of Laser Intensity Data and First Derivative (slope) Map. .

Outcrop Patterns and Structure Contours

A. Refer to Appendix F in back of lab manual for list of commonly used geologic map symbols:

- *Emphasis*:
  - Strike and dip of bedding
- Inclined
- Horizontal
- Vertical
- Overturned
  - Fold axes
- Syncline
- Antincline
- Plunging folds
  - Strike and dip of joints
  - Strike and dip of foliation

- Fault symbols
- Thrust
- Vertical

Outcrop, structure and age relationships:

- Folding of rocks
- Inclined sequence of sedimentary beds
- Layer cake relations
- Oldest on bottom, youngest on top

Exploration efforts are invigorated during this year after 30 years of lull in Kwanza basin onshore. This classic region is in the salt basin of West Africa that covers nearly 25,000 sq.km area. Within the basin, the surface outcrops represent the entirestratigraphic column that ranges from Early Cretaceous to Recent. During the early exploration phase several reservoirs are found in Aptian, Albian and Miocene age group. These are spread in the north portion of the onshore basin. The current objective is to present outcrop study of the key reservoirs in Aptian and Albian.

Classic outcrop sections of prominent reservoirs in Aptian and Albian are illustrated to show the various lithofacies. Clastic,Shaly and Carbonate facies are discernible in Aptian Binga formation. The different facies\ thickness is variable from few millimetres to several metres that represents lagoonal to near shore deposition. The Albian is known under sever al facies. It varies fromanhydritic carbonate (Tuenza), lagoonal (Catumbela), sub-tidal (Oölitic Carbonate) and distal depositional (Quissande) facies that vary from few to hun dreds of metres in thickness.

The petrographic, biostratigraphic and log criteria for Aptian and Albian section further enhance the knowledge on the key reservoir section.To sum up, the study results of the outcrop section for the key reservoirs in Aptian and Albian will be an important understanding to the strata of the subsurface in exploration work.

RELATION TO TOPOGRAPHY

RELATION

Four large, essentially roadless, and relatively

undisturbed areas in the Oklahoma portion of the Ouachita National Forest meet the criteria for wilderness set forth by Public Law 93-622, commonly known as the

Eastern Wilderness Act. Two of the areas have been recommended by the U.S. Forest Service for designation as Wilderness Areas under provisions of the Act. Several organizations concerned with conservation and wildlife protection have recommended inclusion of all four areas in the national wilderness system.

Official designation of an area as wilderness requires action by the U.S. Congress, which has not yet acted on an Oklahoma wilderness bill. The four areas are all in southeastern LeFlore County. Several tree and shrub species reach the western limit of their distribution in the area . Upland forests of the region have been described as oak-pine or oak-hickory-pine (3, 4, 5), but quantitative sampling of the vegetation has not been done.

I undertook the present study in order to investigate the influence of topography on the woody species composition of the forests of the region and to provide a quantitative description of the woody vegetation of the four potential wilderness areas.

STUDY AREA

The four areas under consideration are in the Ouachita National Forest near the Oklahoma-Arkansas border. The sites are in the Ouachita Mountains, which (in southeastern LeFlore County) consist of long linear east-west trending ridges which reach altitudes of 800 m with the intervening valleys at an elevation of about 300 m. Surface rocks are mostly sandstones and shales .

The general area is in the Oak-Pine vegetation type of Duck and Fletcher . The climate is classified as Humid Subtropical, with annual precipitation of 135 cm and mean annual temperature of 17.2 °C.

However, microclimates in mountainous regions vary considerably from the regional climate because of differences in slope, aspect, and elevation . The Beech Creek Area covers

3250 ha and includes most of the watershed of Beech Creek, which is made

up of the floodplain of Beech Creek, the north-facing slope of Walnut Mountain, and the south-facing slopes of Blue Bouncer Mountain and Lynn Mountain. The Upper Kiamichi Area covers 3570 ha and includes most of

Pine Mountain, the source and several kilometres of the upper Kiamichi River, and much of the south-facing slope of Rich Mountain. The Rich Mountain Area covers 2040 ha and includes most of the Oklahoma portion of the north-facing slope of Rich Mountain. The Black Fork Area covers 1775 ha and includes most of the Oklahoma portion of Black Fork Mountain and part of the Big Creek Valley.

METHODS

Site Selection

Because of the east-west orientation of the long ridges in the area, only four topographic types were considered to be important in this study. They are: valley bottom-floodplain, south-facing slope, ridgetop, and north-facing slope. Other topographic types exist, but they cover relatively small areas and wer not sampled. A set of 18 sites was selected in the proposed wilderness areas to obtain at least three sites in each of the four major topographic types. Seven sites were ampled on north-facing slopes because a preliminary survey indicated greater variation in species composition on these sites.

Field Sampling

At each site, 30 sampling points were chosen with a stratified random design in a portion of the site considered to be representative. At each point, trees (defined as woody stems with dbh e″ 5 cm) were sampled for density, frequency, and basal area with an augmented variable-radius method in which basal area was estimated with a two-factor metric forester's prism and trees were counted by species in an 18.3-m diametre circular plot. Woody plants with a dbh < 5 cm (saplings, tree seedlings, shrubs, and woody vines) were counted by species in a 6.1-m diametre circular plot for frequency and density estimates. Numbers of samples and plot sizes were chosen to reduce the standard error of tree density to less than 20% of the mean, on the basis of previous experience in forest sampling.

Data Analysis

Field data were used to calculate standard phytosociological data: density, frequency, and basal area of tree species; and density and frequency of small woody species. Importance values were obtained for each tree species by calculating the mean of the relative density, relative frequency, and relative basal area (expressed as per cent). Importance values for shrubs and vines were calculated as the mean of relative density and relative frequency.

The Shannon-Wiener diversity index and an evenness index were calculated from relative density data for trees. Diversity and evenness were not calculated for shrubs and vines since some were only identified to genus. Reciprocal averaging ordinations were done separately for trees and for shrubs and vines, using importance values. The data were not transformed and minor species were not eliminated or downweighted in the ordination results presented below.

For each site, average slope, aspect, and elevation were determined from 7.5' topographic maps and a relative insolation (RI) index was calculated from the slope and aspect. The RI was defined as the mean ratio of the length of a shadow falling on a slope to the length of the same shadow falling on a horizontal plane, calculated at noon on the solstices and equinoxes. Product-moment correlation coefficients were calculated for elevation and RI with tree density, tree basal area, shrub and vine density, and ordination axis positions.

RESULTS AND DISCUSSION

The ordination based on tree species, which was derived entirely from quantitative vegetation data, resulted in a grouping of the sites that is consistent with the topographic positions of the stands. It also confirms subjective impressions formed from field observations. The ordination placed sixteen of the sites into four distinct groups: ridgetop (TOP), upper and middle north-facing slope (UMN), valley bottom-floodplain (BOT), and south-facing slope (S). The other two sites were from lower north-facing slopes (LN) and were placed between the BOT and S groups. The rdination based on shrub and vine species, which is not shown, resulted in the same grouping of sites as the tree ordination.

The first axis of the ordination based on tree species had a product-moment correlation of -0.761 ($p < 0.01$) with the relative insolation (RI) index and the second axis had a correlation of -0.789 ($p < 0.01$) with elevation.

In the ordination based on shrubs and vines, the first axis had a correlation of -0.807 with RI and the second axis had a correlation of 0.620 with elevation. A list of the 37 tree species encountered in the sample plots and their importance per centages in the five community types is given in Table. A similar list of the 19 species/genera of shrubs and woody vines is given in Table. Latin names correspond to those given by Waterfall, except for the two species of *Tilia*, which were determined from a key given by Lawson . Most of the tree species encountered are easily identified without a close, time-consuming examination.

However, the linden (*Tilia*) species and the northern red oak-Shumard's oak (*Quercus rubra-Q. shumardii*) pair present some difficulty when many trees are encountered. Of the lindens that were closely examined, *Tilia americana* was found only in the valley bottom-floodplain community and *T. caroliniana* on the north-facing slopes and ridgetops. None of the oak trees which were closely examined appeared to be northern red oak, although Palmer reported that both northern red oak and Shumard's oak were common on Rich Mountain.

Although many of the tree species occurred in all five community types, most species reached a definite maximum in one type. Species which were found in more or less equal numbers in the five types are dogwood (*Cornus florida*), serviceberry (*Amelanchier arborea*), red maple (*Acer rubrum*) American elm (*Ulmus americana*), and black cherry (*Prunus serotina*). The valley bottom-floodplain community (BOT) had 24 species, with no species being dominant (importance per centage e" 20).

This community offers a diversity of habitats, from creek margins and sloughs to gentle but rocky and well-drained slopes, which is reflected in the diversity of vegetation. Several tree species are important in most stands, including sugar maple (*Acer saccharum*), bitternut hickory (*Carya cordiformis*), sweet gum (*Liquidambar styraciflua*), ironwood (*Ostrya virginiana*), Shumard's oak, and white oak (*Q. alba*). A few large shortleaf pines (*Pinus echinata*) are found scattered throughout this type. American beech (*Fagus grandifolia*), which reaches the western limit of its distribution in eastern LeFlore County, is common along Beech Creek and a few individuals can be found along the upper Kiamichi River and Big Creek. Shrubs and vines are abundant (H" 5500/ha), the most important of which are briers (*Smilax* spp.) and poison ivy (*Rhus radicans*).

The forest community found on upper and middle north-facing slopes (UMN) has many tree species, but species diversity is much lower than in the BOT community because of the strong dominance of mockernut

(*Carya tomentosa*), which has an importance of over 50% in some stands. The only other tree species with importance greater than 10% are linden (*Tilia caroliniana*) and white oak, although 10 species have their maximum importance in this type (Table). Another tree species that reaches the western limit of its distribution in this area is cucumber tree (*Magnolia acuminata*), which is abundant at some sites on the upper north-facing slope of Rich Mountain but absent from very similar sites on the north side of Black Fork Mountain. Pine is absent and the oaks reach their lowest importance in this type. The most important shrubs and vines are coralberry (*Symphoricarpos orbiculatus*), Virginia creeper (*Parthenocissus quinquefolia*), and poison ivy.

The sites on lower north-facing slopes (LN) have tree vegetation that is somewhat intermediate to that of the BOT and S types. The most important tree species is Shumard's oak, which also has a higher importance value here than in any of the other types. Other important tree species are mockernut, red maple (*Acer rubrum*), black gum (*Nyssa sylvatica*), pine, and white oak. The most important shrub and vine species in this type are poison ivy, brier, witch hazel (*Hamamelis virginiana*), and sugar uckleberry (*Vaccinium vacillans*). The vegetation of south-facing slopes (S) is typical oak-pine forest, with shortleaf pine accounting for about 30% of the tree

DIP AND STRIKE OF BED

STRUCTURTAL GEOLOGY

Structural geology is the branch of geology that deals with the three-dimensional shapes of rocks and how these shapes form. These shapes, or structures, develop when rocks are deformed as a result of some force applied to the rock. The force that acts upon a rock is the stress. As a result of the applied stress, the rock will deform; that is, it will undergo some type of change in shape or volume. This deformation is called the strain. The three-dimensional shape that results is called a geologic structure.

STRIKE AND DIP

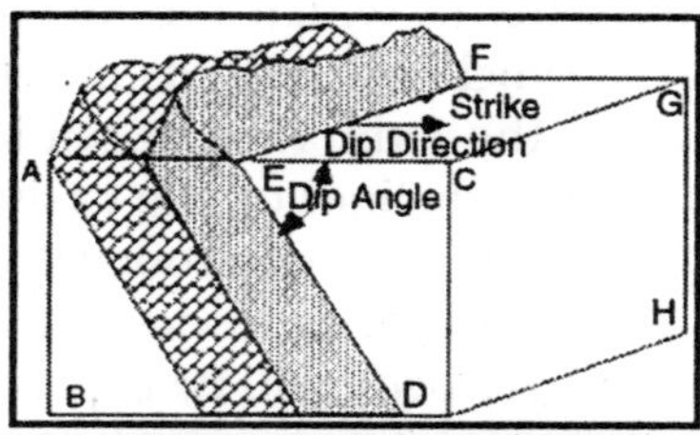

Folds and faults are three dimensional structures that may have various orientations in space. We must have some method of describing the orientation of these structures on a geologic map. The method that geologists use to describe the location in space of a geologic structure is the strike and dip. The strike of a bed or structure is given by the compass direction of an imaginary line at the intersection of a horizontal plane and the inclined bed or structure. The dip of a bed or structure is the angle of intersection between a horizontal plane and the inclined bed or structure. Dip is always measured perpendicular to strike. The dip determination will also have a direction, i.e., that direction toward which the layer is inclined down from the horizontal plane.

FOLDS

If a rock undergoes compressional stress and plastic strain, the resultant geologic feature will be a fold. Most folds develop at deeper levels within the crust, where rocks are under a high enough temperature and pressure to deform plastically. Folds are defined by the relationship of the dipping layers to an imaginary plane that divides the fold in half. This plane is called the axial plane. The line of intersection between the axial plane of a fold and the fold is called the axis.

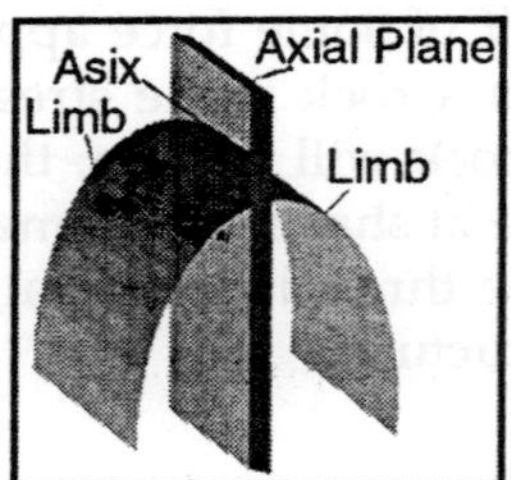

Anticlines are folds in which the limbs of the fold dip away from the axis. Synclines are folds in which the limbs of the fold dip toward the axis. It is possible to distinguish between anticlines and synclines on the basis of the age of the bed exposed in the middle of the eroded fold. In an eroded anticline, older beds will be exposed in the middle; in an eroded syncline, younger beds will be exposed in the middle.

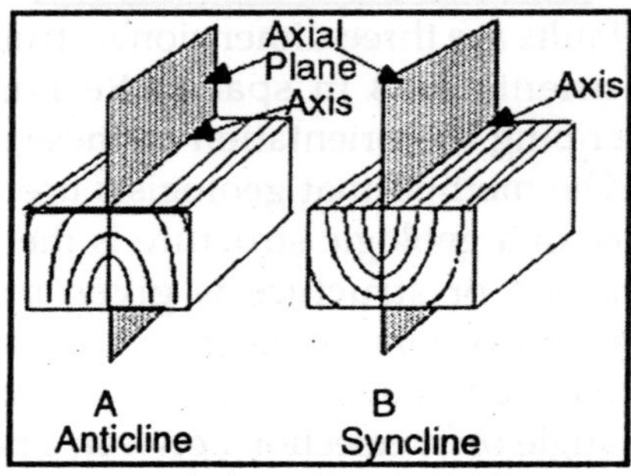

FAULTS

Faults occur when a rock is subjected to enough stress that it reaches the limit of its strength and breaks. Different types of faults will form depending on the type of stress to which the rock is subjected.

Strike-slip faults develop when rocks are subjected to shear stress. In strike-slip faults, the movement is purely horizontal, with no up-and-down displacement. We classify strike-slip faults as either right lateral or left lateral. Imagine yourself standing on one side of a strike-slip fault, so that you are facing the fault.

If the block on the other side is displaced to your right, the fault is a right lateral strike-slip fault . If the block on the other side of the fault is displaced to your left, the fault is a left lateral strike-slip fault.

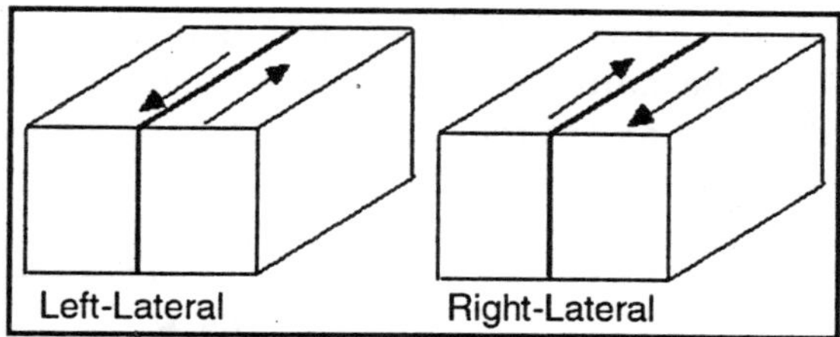

Dip slip faults form when rocks break from compressional or tensional stress, so that there is vertical displacement. When distinguishing between the types of dip slip faults, we define the block of rock above the fault plane as the hanging wall block and the block of rock below the fault plane as the footwall block. The terms "hanging wall" and "footwall" are old mining terms. The hanging wall block was the block of rock where miners would hang their lanterns; the footwall block was where miners would walk.

It is easy to determine which side of the fault is the hanging wall if you imagine a miner standing on the fault plane. The hanging wall will be the block above the miner1s head; the footwall will be the block below the miner1s feet.

If rocks break under tensional stress, the hanging wall will move down relative to the footwall and a normal fault forms'.

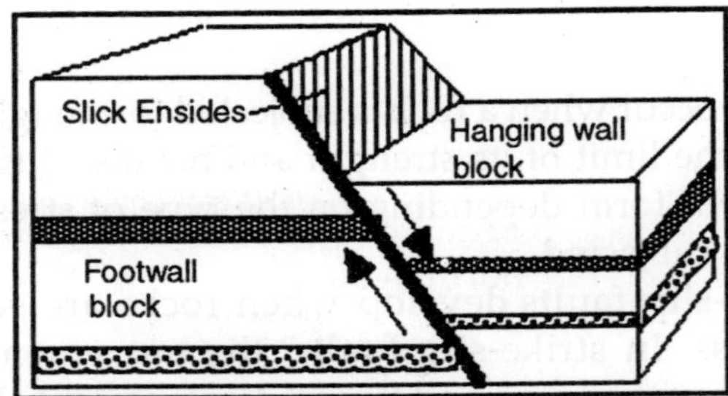

In this situation, the crust actually extends and lengthens. When rocks break under compressional stress, the hanging wall moves up relative to the footwall, and a reverse fault or thrust faultforms. In a reverse or thrust fault, the crust is shortened. Thrust faults are simply reverse faults in which the angle formed by the fault plane and the surface is quite shallow.

Commonly, the fault plane itself can become grooved and polished as one block of rock scrapes against the other. The scratches on the fault plane surface are called slickensides. Slickensides may record the slip orientation of the fault plane, and may even feel smoother in the direction of slip.

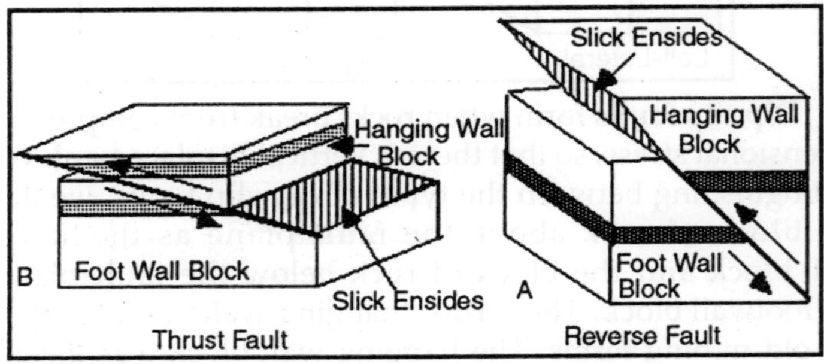

EXERCISES

- Use the written or symbolic form of the strike and dip measurement to express the following orientations. Use a protractor to determine strike. For the problems in this exercise, assume that north is at the top of the page unless otherwise indicated.
  - *Strike*: north thirty-seven degrees west;

- *Dip*: Twenty-five degrees southwest

- *Strike*: North eighty-nine degrees east, Dip: five degrees northwest

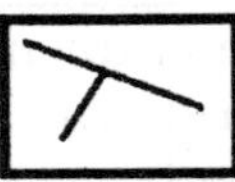

- Determine if the fault below is a right or left lateral strike-slip fault.

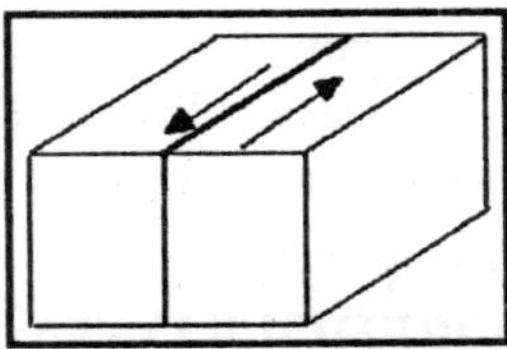

- Name the structures shown in the figures below.

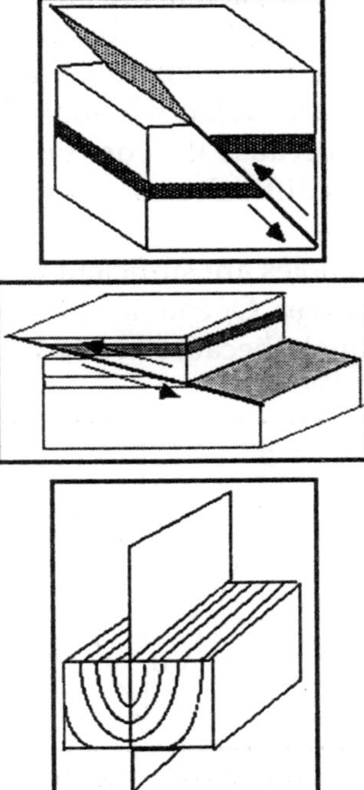

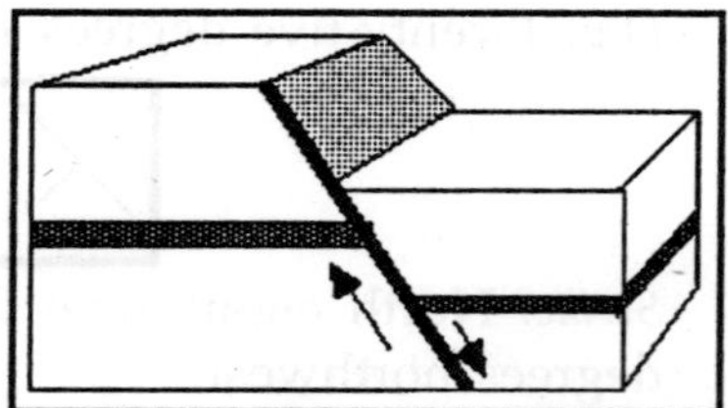

- Use the structure models to answer the following questions. You will have to print the models out to complete the lab.
  - Name the structure.
  - Label the oldest to youngest beds X-Y-Z.
  - Complete the sides.
  - Draw a line indicating the fold axis on the model when appropriate.

STRIKE AND DIP FROM VERTICAL BOREHOLE DATA

When dipping layers are encountered in a borehole, and we have a core sample to study, we know two facts: the dip of the bed relative to the borehole, and the location of the bed below the surface. What we don't know is the strike and the dip direction, because the core sample rotates as it comes out of the hole. It is difficult and expensive to get oriented samples from drill holes.

Two boreholes are sufficient to narrow the strike and dip to two alternatives. Of course, three boreholes define the strike and dip uniquely because we have three elevation points in the plane.

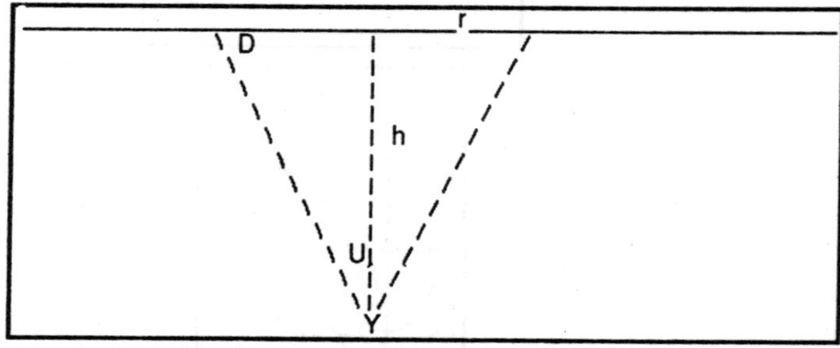

In the borehole above, the bedding plane has dip D and

intersects the core with angle U. The bed could have all possible strikes, and the range of possible bedding orientations outlines a cone with an apical angle of 2U.

If the angle U is significantly different in the two boreholes, the structure is not planar.

If the sample occurs at depth h, the possible outcrops of the bed at the surface lie somewhere on a circle of radius r. The radius of the circle is given by r = h Cot D = h Tan U. (You can also solve the problem graphically, by drawing a cross-section as above)

Two such boreholes define two circles. In practice it is most convenient to refer depths to a common datum elevation. When we do this, the strike of the plane must be tangent to both circles. This construction requires that the circles be on the same plane. If they're not, then a line tangent to the circles won't be horizontal, and therefore will not be the true strike. Thus all depths have to be calculated with respect to some convenient datum plane.

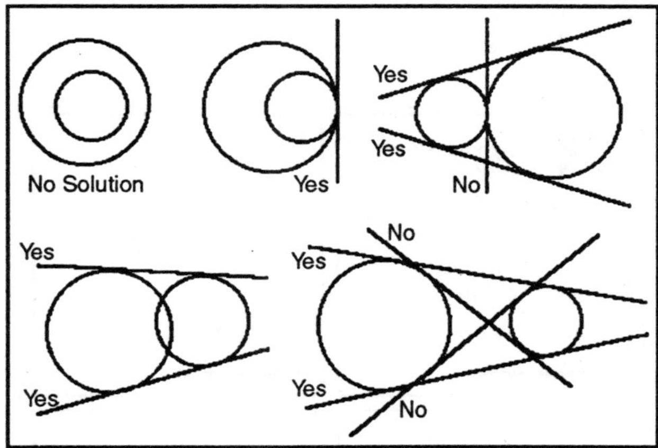

If one circle contains the other, there are no solutions. This most often happens if there is a fault or other structural disturbance between the two boreholes. If the circles intersect there are only two lines tangent to both and both are possible strike directions. If the circles touch, there are three tangents and if the circles are separate there are four. However, in every case only two are possible strike directions.

The rule for telling valid strike directions is simple: lines that pass between the circles are not permissible because the bed would have to dip in two directions at once.

In practice, and why this technique is useful, is that we can often rule out one of the two strike directions because it does not agree with other geologic data from the area. If we know the structures are upright folds trending 25 degrees, and our strike directions from borehole data are 27 and 84 degrees, the 27 degree strike is much more likely to be correct.

IMPORTANT SPECIAL CASES

We may have two boreholes but only one dip measurement. For example, the critical portion of the core may be too highly farctured in one borehole. Two boreholes allow us to confirm that the structure has the same dip in both and is in fact a plane. If we only have one dip determination, we have to assume the structure is a plane (which may or may not be true). Assume the dip is the same in both the known and the unknown borehole and proceed as usual, but be careful in applying the results.

Example:

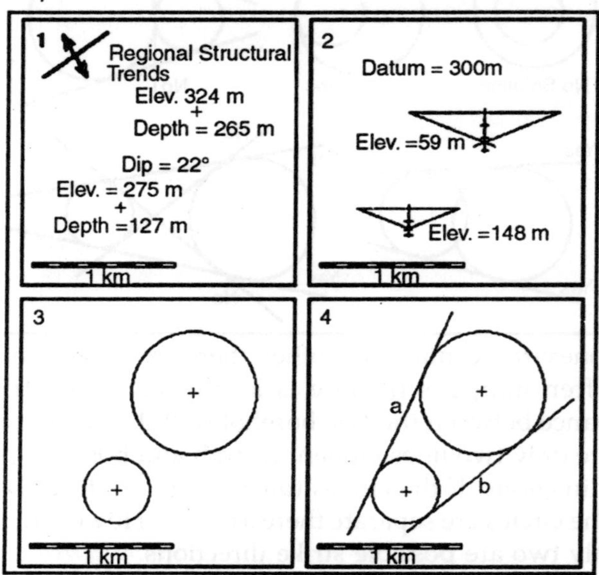

1 Given the borehole data, find the possible strike of the marker bed encountered in each borehole. We also know that regional structural trends are as shown.
2 By construction or calculation, find the radii of the circles where the bedding plane might intersect the surface.
3 Construct the circles.
4 Draw the possible strike orientations (you already know the dip). In this case direction b is most likely to be the strike direction.

It may happen that we have a borehole and an outcrop, but we cannot get an attitude measurement (too massive, too fractured, etc.). We can think of the outcrop as a borehole of zero depth. If we place the datum plane at the elevation of the outcrop, we will have one circle defined by the borehole and a point (or a circle of zero radius) at the outcrop. The strike must pass through the outcrop and be tangent to the circle.

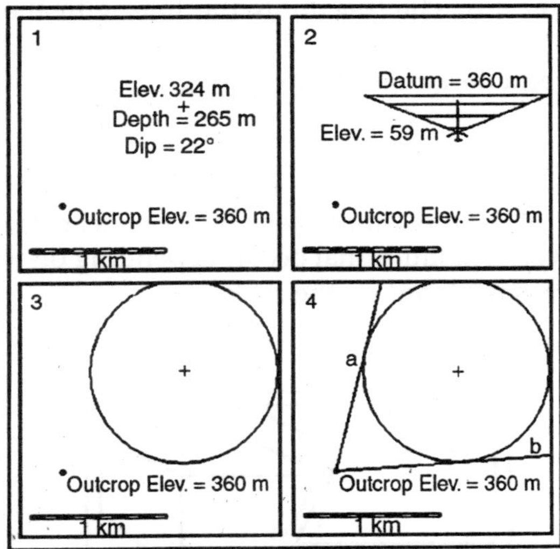

1 We have a single borehole and an outcrop.
2 By construction or calculation, find the radius of the circle where the bedding plane might intersect the datum plane at the elevation of the outcrop.

3 Construct the circle.
4 Draw the possible strike orientations.

MEASURING STRIKE AND DIP

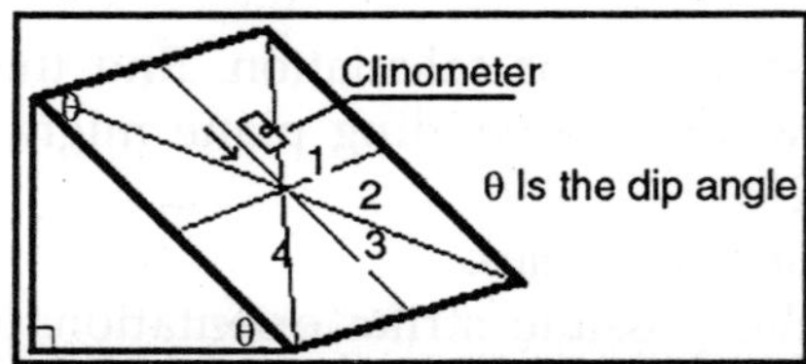

Strike and dip are measurements that are needed to define the orientation of a plane. The *strike line* runs horizontal to the plane. It is measured in degrees or bearing. In quadrant compasses, it is recorded in terms of north. (N72E). Azimuth compasses can read from 0 degrees to 360 degrees. The *plunge* is the direction in which the plane runs (the way it tilts). It is measured in respect to its position from the horizontal. All lines on the plane are plunging, *except* the strike line. Lines 2 and 4 are referred to as the *apparent dip,* and are always less than the true dip. The *true dip* is the line with the steepest inclination. It is expressed in degrees and direction. Note that the dip line and the strike line are 90 degrees from one another.

Faults

When stress builds up in a rock, it will break. Faults are the product of compressional, tensional or shear stress. There are several types of faults that occur, depending on where and how the stress is placed on the rock.

Some Common Faults

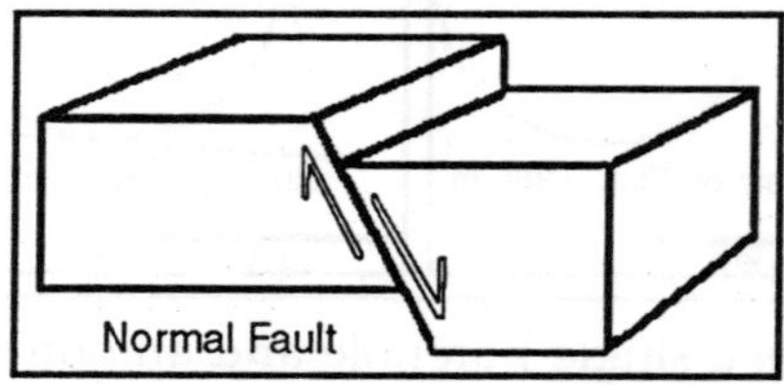
Normal Fault

Dip-slip faults: Can be formed from either compressional or tensional stress. Normal faults are caused by tensional forces and reverse faults are caused by compressional forces.

Strike-slip Faults: Direction of movement is parallel to the strike. If an observer stands on one side of the fault and sees a person move with the fault on the other side, depending on the direction that person appears to move, will be the classification of the fault

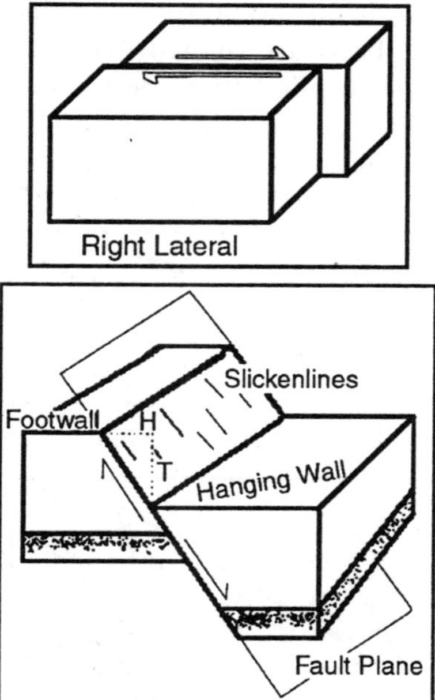

Dip-slip Normal Fault: The fault shown in the above diagram is a *normal fault*. The break occurred due to tensional forces. The hanging wall moves down in relation to the footwall. It is considered a "high angle" fault, with an angle usually between 45 and 90 degrees. A fault has a strike, dip and movement. The movement is measured by the *heave* (H on diagram), representing the horizontal distance moved and the *throw* (T on diagram), which represents the vertical distance moved. *Slickenlines*are fine "scratch" lines present on the fault surface that indicates the direction of the slip.

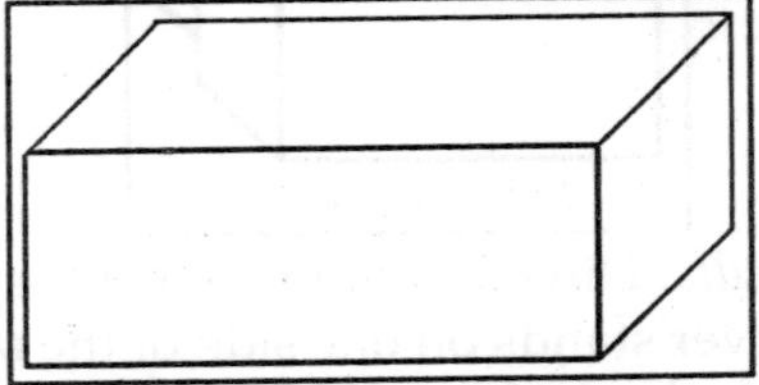

Folds

Sediment is deposited in horizontal layers, called "beds". The oldest sediment is on the bottom and the youngest is on the top (unless the beds are overturned). After deposition, compressional stress applied to a rock will cause it to fold. There are two main types of folds. An *anticline* is a fold that bends downward, creating a hill-like structure. A *syncline* is a fold that bends upward in the shape of a "U".

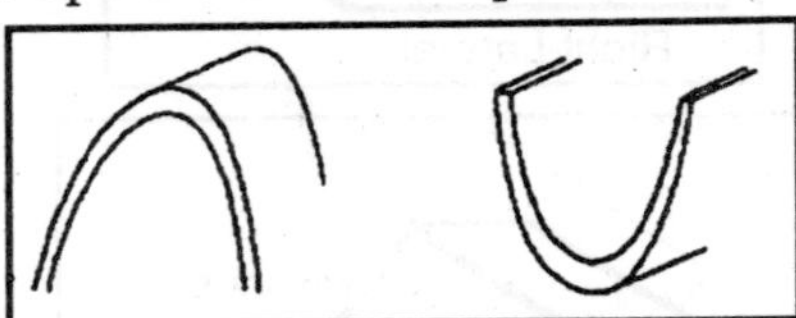

Fig. Anticline Syncline

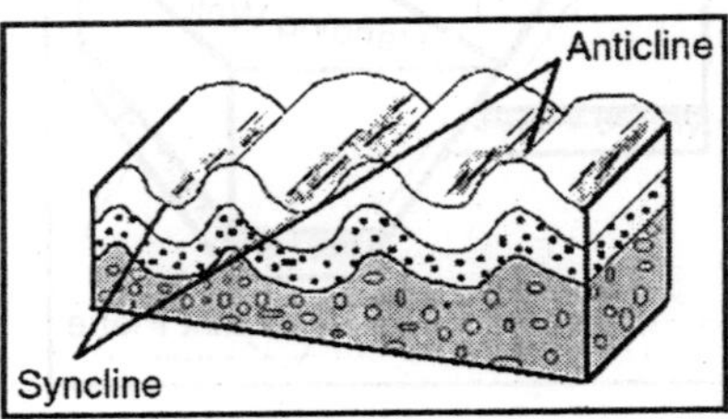

Folds are considered either *cylindrical* or *non-cylindrical.* If you were to slice a cylindrical fold (like bread), the sections would appear similar.

If you could to take a pencil, without lifting it from the surface of the fold, and bring it from one limb to the other, the fold is cylindrical. Looking at the diagram of the non-cylindrical fold, it is evident that the pencil could not remain on the surface

Fig. Cylindrical Non-Cylindrical

Fig. Concept of Cylindrical Fold

The *plunge* of a structure is measured as an angle with respect to its position from the horizontal. Folds can be plunging or non-plunging, according to the inclination of its fold axis.

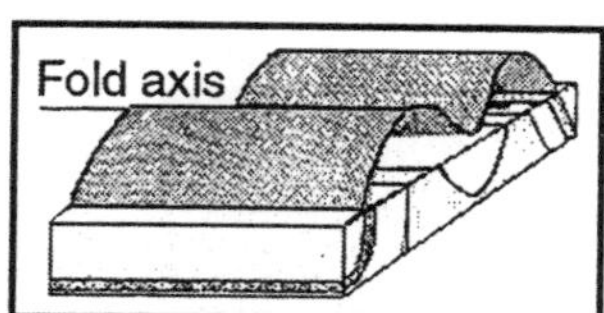

Fig. Non-Plunging Fold

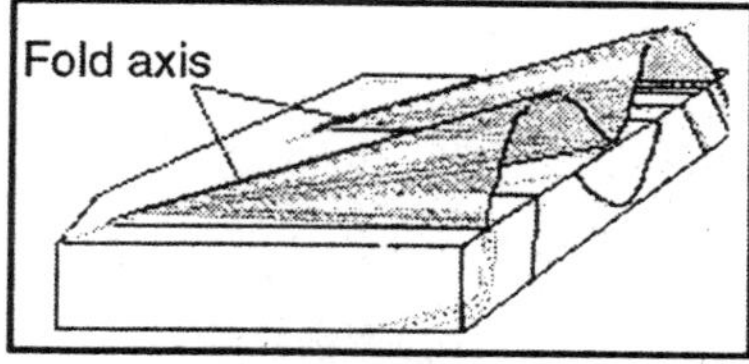

Fig. Plunging Fold

The above photograph represents a fold that occurs in Unit 2. Strikes and dips were taken on both limbs and the data was plotted on a stereonet in order to determine the orientation of the fold axis.

Using Stereographic Net

A stereograph is used by geologists to plot the strike and dip of folds, faults and bed layers uncovered in the field. A *Wulff Net* stereograph consists of divided areas of equal angle. This will be the type of stereonet used for demonstrations shown here.

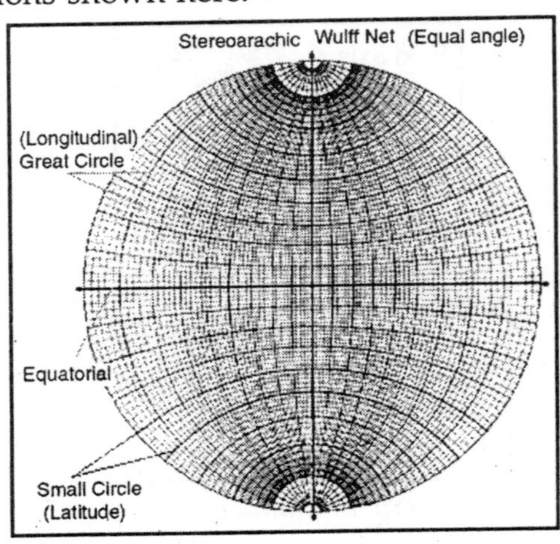

All longitudinal lines and the equatorial line are considered *Great Circles*. A plane that intersects the sphere and passes through the centre, lies on a Great Circle. All latitude lines (except the equatorial line) are *Small Circles*. A plane that intersects the sphere *without* passing through the centre, lies on a Small Circle. Lines are divided into 2 degree intervals, with the darker lines representing 10 degree marks.

Preparing the Stereonet

Place a thumb tack through the the centre of the stereonet. Next, place a sheet of tracing paper over the net, pushing through the tack. Put a small pencil eraser on top of the tack (to avoid injuries). Trace over the circumference of the net and mark "North" with a tick mark and "N" on the tracing paper.

Plotting a Line: (Coordinate: (120,20SE)

- Count 120 degrees (clockwise) east from North. Put a tick mark on this spot.
- Turn the tracing paper until this mark aligns with the E-W equatorial Great Circle.
- From the tick mark, count toward the centre 20 degrees. Put a dot on this spot.

Plotting a Plane: (Coordinate: (040,45SE)

- Count 40 degrees (clockwise) east from North. Put a tick mark on this spot.
- Turn the tracing paper so the "40 degree tick mark" is aligned with North.
- From the East end of the equatorial line, count in 45 degrees. Mark with a dot.
- With the dot lined up on one of the longitudinal Great Circles, trace along the Great Circle from top to bottom. Rotate the tracing paper back where both "North's" line up to view the orientation of the plane.

Plotting a Fold: (Coordinate: (020,40W) and (325,70NE)

- Count 200 degrees (clockwise) from North. Put a tick mark on this spot.

- Turn the tracing paper so the "200 degree tick mark" is aligned with North.
- From the East end of the equatorial line, count in 40 degrees. Mark with a dot.
- With the dot lined up on one of the longitudinal Great Circles, trace along the Great Circle from top to bottom.
- Repeat the same procedures for the second coordinate.

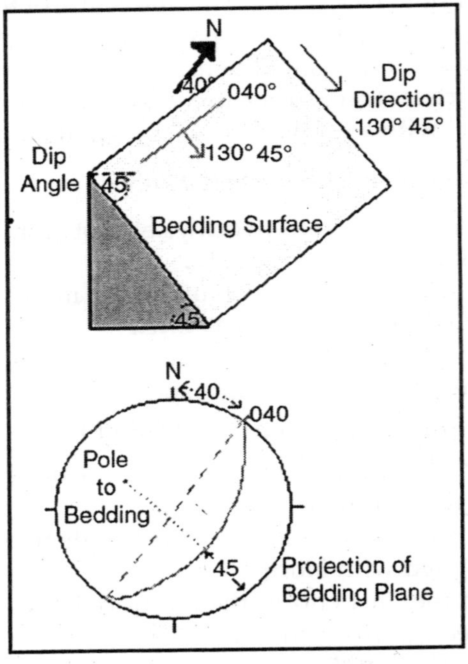

Plotting a Plane

A bedding plane is measured in the field and the following coordinate is established: (040,45) The strike of the plane is N40E and the plane dips 45 degrees SE. The dip *direction* is N130E.

To represent this plane on a stereonet, count (clockwise) 40 degrees east from North. The red dashed line represents the strike of the plane. The solid red line represents the dip of the plane. All points on this line are the apparent dip, other than the 45 degree mark (which is considered the true dip). The pole to the plane is "normal" to the plane and can be located on the stereonet by counting 90 degrees from the "true dip" mark on the solid red line. See the above instructions on how to accurately plot these points on a stereonet.

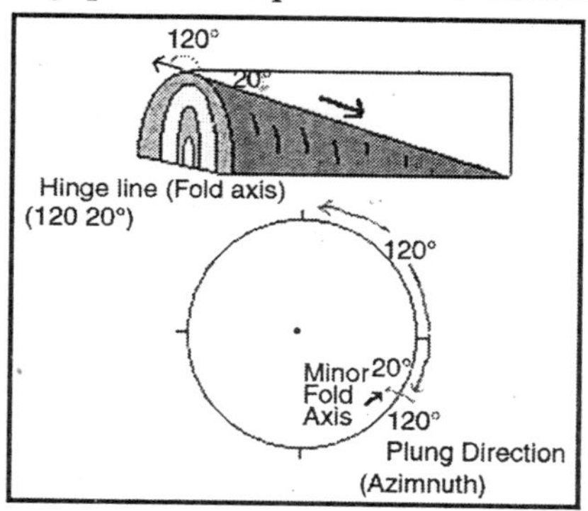

Plotting a Fold Axis

A fold axis is measured in the field with the following coordinate: (120,20). The strike of the fold axis is N120E and it is dipping 20 degrees SE. The fold axis is represented on the stereonet by counting (clockwise) 120 degrees east of North. (red "tick mark") The red dot represents the 20 degree dip in the fold axis.

Note: A fold axis is a "line", which is represented on a stereonet as a dot

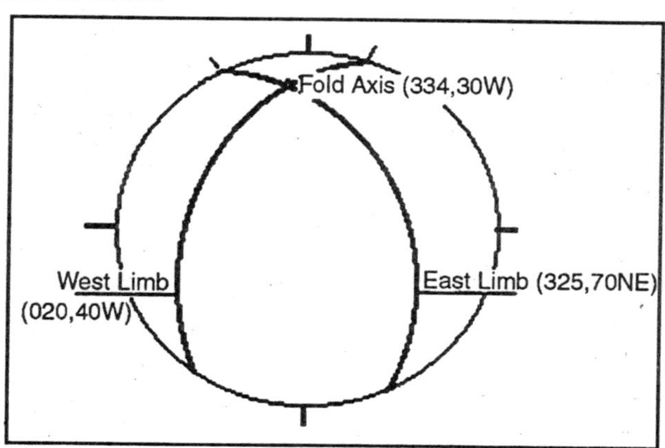

Plotting a Fold and Finding its Axis

A fold is measured in the field by obtaining the strike and dip of each of the limbs. The fold represented on the stereonet to the left has one limb orientated at 325 degrees and dips 70 degrees to the NE. The other limb is orientated at 334 degrees and dips 30 degrees west.

The fold axis is a line that represents maximum curvature of the fold and separates the two limbs. On a stereonet, this can be found where the two limbs cross each other. In this case, it would be at: (334,30W).

OVERLAP

DEFINITION

When a .series of strata is deposited in a basin with sloping sides, or one sloping side, each bed will extend farther than the one upon which it lies, and thus in a thick mass of strata, if the shelving bottom be gently inclined, the upper beds will extend far beyond the lower ones, or overlap them . Overlap also occurs where the sea is advancing or transgressing slowly across a subsiding land surface, the rate of depression not much exceeding the rate of deposition. Here also each stratum extends farther across the old land surface than the one beneath it, and conceals the edges of the latter. The relation of overlap is between the successive layers of a conformable series.

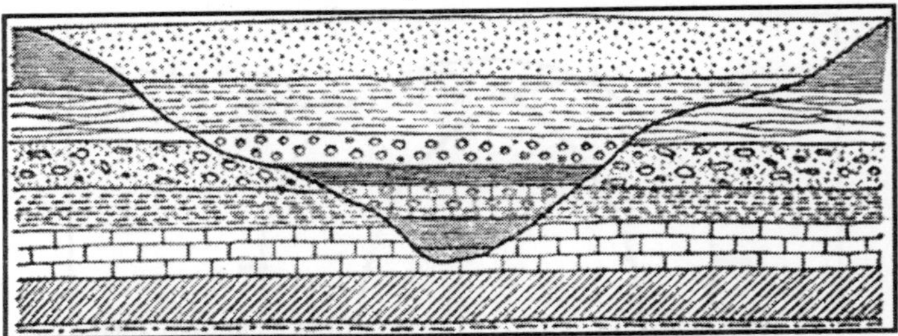

Fig. Unconformity Without Change of Dip, and Overlap.

Overlap may be a structure of much economic importance, if one of the lower strata, say a coal-bed, is mined. It is not safe to assume that wherever the upper beds of such a series are found, the lower will be found directly beneath them, an assumption which may result in costly failure.

OUTLIER AND INLIER

INLIERS/OUTLIERS

Not every outcrop problem produces a simple, neat line on the map. Very often there are disconnected areas of outcrop

An *Outlier* is an outcrop outside the main exposure area. An *Inlier* is an outcrop within the main exposure area, exposed through a window in the overlying rocks.

Any time you have a structure or topographic contour that blocks you from connecting two outcrop areas, you are probably dealing with outliers or inliers (at least within the map area — they may connect outside the map area).

Always be prepared for outliers near hilltops and inliers near valley bottoms. Any time new points suddenly appear on a structure contour, be prepared for a possible inlier or outlier.

Any time your outcrop band shows deep re-entrants, you may have outliers or inliers. The available evidence may not be sufficient to tell for sure. Sometimes even a thin layer will coincide with the land surface and outcrop as an area rather than a band. The clue that this may be happening is a structure contour that coincides closely with a topographic contour.

INLIER, OUTLIER

The terms 'inlier' and 'outlier' each describe a situation where a outcrop of one type of rock appears at the surface completely surrounded by another. This is usually a result of erosion having removed part of an overlying rock unit from an underlying unit. An inlier is then an isolated region of underlying rock exposed by erosion of the overlying units, whereas an outlier is an isolated remnant of overlying rock that has survived the erosion that has stripped it away everywhere else to reveal the underlying rocks.

In regions of flat-lying or gently dipping strata, inliers tend to occur in valley bottoms whereas the hilltops may be occupied by outliers. Where the strata have been strongly folded prior to erosion these simple topographic relationships do not necessarily apply, and it is possible to have an inlier preserved in the core of an anticline occupying high ground or an outlier preserved in the core of a syncline occupying low ground.

Inliers and outliers can also be generated by fault movements followed by erosion. For example, if an older unit is thrust over a younger unit erosion can isolate part of the thrust sheet, leaving an outlier of older rock surrounded by younger rock (a reversal of the usual age relation-ship).

Similarly erosion can expose a window through the thrust sheet to reveal an inlier of younger rock surrounded by older rock. Fault movements by which blocks of terrain move up or down relative to each other, as when horsts or grabens are formed, can also produce inliers and outlier

Example

This is exactly the same map used to illustrate outcrop patterns in areas of complex topography. The only difference is that the structure contours are 100 metres higher.

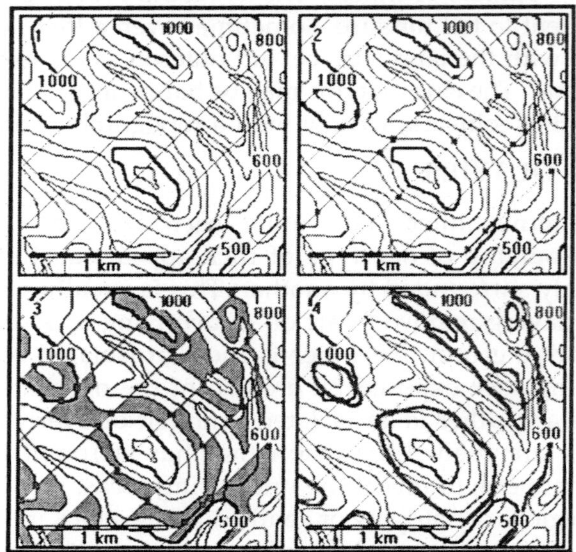

- Find the outcrop pattern of the coal seam whose structure contours are shown in red.
- Locate intersections of like contours.
- Intervals where the outcrop can be are shown in purple. Topography is higher than the coal seam in yellow areas and lower in green. Note some purple areas are completely disconnected.
- Draw the outcrop band. Start with simple areas, locate obvious ridge and valley crossings, then finish difficult areas.

OUTLIER DETECTION AND INLIER FUSION

As measurement noise we assume a contaminated Gaussian distribution with a main peak within a small interval (of 1 pixel) and a small per centage of outliers. Inlier noise is caused by the limited resolution of the disparity matcher and by the interpolation artifacts.

Outliers are undetected correspondence failures and may be arbitrarily large. As threshold to detect the outliers we utilize the depth uncertainty interval ^{ek}. The detection of an outlier at k terminates the linking at k–1. All depth values [d_k, $d_{k+1}...d_{1-1}$] are inlier depth values that fall within the uncertainty interval around the mean depth estimate. They are fused by a simple 1-D kalman filter to obtain an optimal mean depth estimate.

Figure (right) explains the outlier selection and link termination for the up-link. The outlier detection scheme is not optimal since it relies on the position of the outlier in the chain.

Valid correspondences behind the outlier are not considered anymore. It will, however, always be as good as a single estimate and in general superior to it. In addition, since we process bidirectionally up- and down-link, we always have two correspondence chains to fuse which allows for one outlier per chain.

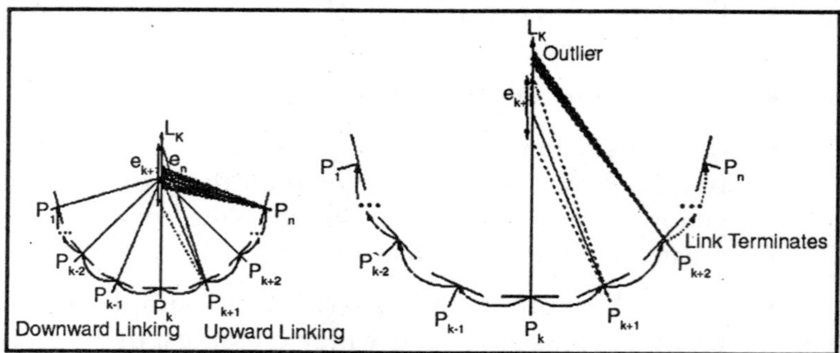

Fig. Depth Fusion and Uncertainty Reduction.

COMMENTS

The colouring scheme used in is useful in extremely complex areas to help keep track of where the outcrop can occur. It shows that the topographic and structure contours intersect to form boxes, often very irregular ones, and that the outcrop band enters at the lower corner and exits at the higher one. Any time you see a purple band bounded on one side by green and on the other by yellow, the outcrop must pass between the contours. The reason is that the topography spans an entire contour interval and therefore mustintersect the bed somewhere in that interval. This is fairly obvious when the two pairs of contours enclose a fairly neat box, but it applies equally well in all situations. It tells you, for example, that the outcrop must pass around an isolated hilltop, for example.

Saddles are inherently ambiguous. In the northeast corner a small hilltop is shown as surrounded by an outcrop band. The colouring shows us that the topography in that area spans an entire contour interval, therefore it must intersect the bed somewhere. Thus the outcrop must pass around the outside of the hill. What cannot be determined is whether the outcrop is connected to the main outcrop band, disconnected as shown, or perhaps just barely connected. There's no right or wrong answer in many cases, we just have to examine the data and draw the most likely configuration.

The saddle in the left-centre of the map is shown as separating two outcrop areas, but in fact they could connect.

There are three possibilities: the outcrops could form two separate loops as shown, the outcrop could enclose the pair of hills in a peanut-shaped curve, or the areas could just barely touch in the saddle to form a . This happens when the saddle just barely grazes the rock unit.

If the saddle is deep and wide, the chances of the outcrop passing through the saddle are greater. If the saddle is shallow and narrow, they are less. The closer the saddle is to the upper structure contour, the more likely it is that the outcrop passes through the saddle. Since the saddle in the left-central part of the map is wide and neatly straddles the structure contour interval, the outcrop is drawn as two separate loops.

In the extreme southwest corner there are two disconnected outcrop areas. The colouring in shows they cannot connect directly. However, we see from the map colouring that the map is incomplete. The outcrop band must also pass down the purple band just left of the scale bar. The 900-metre topographic and structure contours appear to intersect again just outside the map area. The upper outcrop band should pass through that intersection and turn south to re-enter the map area. It could connect to the lower band somewhere outside the map area.

Chapter 4

Rock Deformation

EARTH ROCK

Within the Earth rocks are continually being subjected to forces that tend to bend them, twist them, or fracture them. When rocks bend, twist or fracture we say that they deform (change shape or size).

The forces that cause deformation of rock are referred to as stresses (Force/unit area). So, to understand rock deformation we must first explore these forces or stresses.

STRESS AND STRAIN

Stress is a force applied over an area. One type of stress that we are all used to is a uniform stress, called pressure. A uniform stress is a stress wherein the forces act equally from all directions.

In the Earth the pressure due to the weight of overlying rocks is a uniform stress, and is sometimes referred to as confining stress.

If stress is not equal from all directions then we say that the stress is a differential stress.

Three kinds of differential stress occur.

1. *Tensional stress (or extensional stress),* which stretches rock;
2. *Compressional stress,* which squeezes rock; and
3. *Shear stress,* which result in slippage and translation.

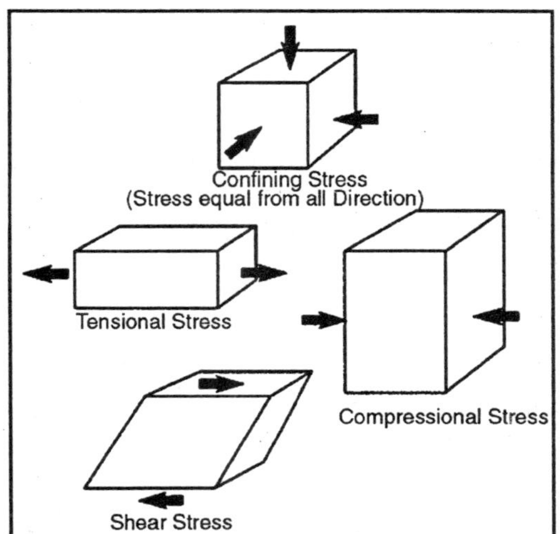

When rocks deform they are said to *strain*. A strain is a change in size, shape, or volume of a material.

STAGES OF DEFORMATION

When a rock is subjected to increasing stress it passes through 3 successive stages of deformation.

Elastic Deformation

Wherein the strain is reversible.

Ductile Deformation

Wherein the strain is irreversible.

Fracture

Iirreversible strain wherein the material breaks.

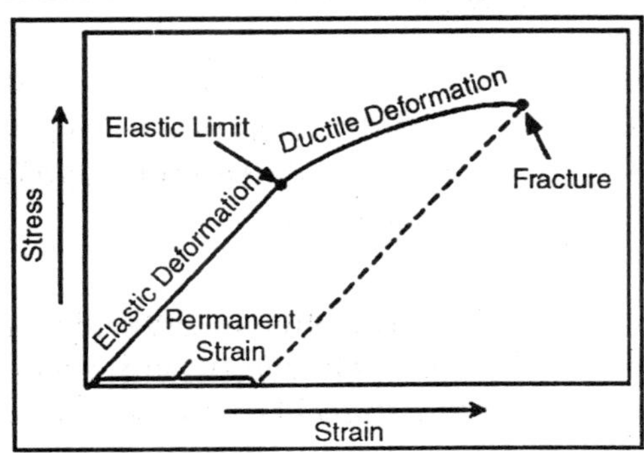

We can divide materials into two classes that depend on their relative behaviour under stress:

- Brittle materials have a small or large region of elastic behaviour but only a small region of ductile behaviour before they fracture.
- Ductile materials have a small region of elastic behaviour and a large region of ductile behaviour before they fracture.

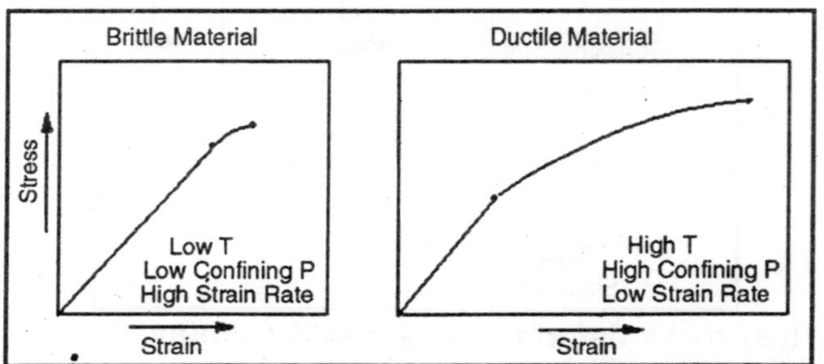

How a material behaves will depend on several factors. Among them are:

Temperature

At high temperature molecules and their bonds can stretch and move, thus materials will behave in more ductile manner.

At low Temperature, materials are:

- Brittle. *Confining Pressure*: At high confining pressure materials are less likely to fracture
- because the pressure of the surroundings tends to hinder the formation of fractures. At low confining stress, material will be brittle and tend to fracture sooner.
- *Strain rate*: At high strain rates material tends to fracture. At low strain rates more time is available for individual atoms to move and therefore ductile behaviour is favoured.

Composition

Some minerals, like quartz, olivine, and feldspars are very brittle. Others, like clay minerals, micas, and calcite are more ductile This is due to the chemical bond types that hold them together. Thus, the mineralogical composition of the rock will be a factor in determining the deformational behaviour of the rock. Another aspect is presence or absence of water. Water appears to weaken the chemical bonds and forms films around mineral grains along which slippage can take place. Thus wet rock tends to behave in ductile manner, while dry rocks tend to behave in brittle manner.

BRITTLE-DUCTILE PROPERTIES

We all know that rocks near the surface of the Earth behave in a brittle manner. Crustal rocks are composed of minerals like quartz and feldspar which have high strength, particularly at low pressure and temperature. As we go deeper in the Earth the strength of these rocks initially increases. At a depth of about 15 km we reach a point called the brittle-ductile transition zone. Below this point rock strength decreases because fractures become closed and the temperature is higher, making the rocks behave in a ductile manner. At the base of the crust the rock type changes to peridotite which is rich in olivine. Olivine is stronger than the minerals that make up most crustal rocks, so the upper part of the mantle is again strong. But, just as in the crust, increasing temperature eventually predominates and at a depth of about 40 km the brittle-ductile transition zone in the mantle occurs. Below this point rocks behave in an increasingly ductile manner.

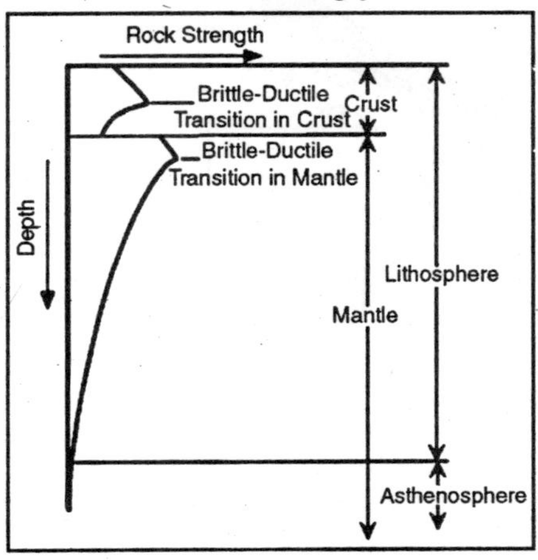

DEFORMATION IN PROGRESS

Only in a few cases does deformation of rocks occur at a rate that is observable on human time scales. Abrupt deformation along faults, usually associated with earthquakes caused by the fracture of rocks occurs on a time scale of minutes or seconds. Gradual deformation along faults or in areas of uplift or subsidence can be measured over periods of months to years with sensitive measuring instruments.

EVIDENCE OF FORMER DEFORMATION

Evidence of deformation that has occurred in the past is very evident in crustal rocks. For example, sedimentary strata and lava flows generally follow the law of original horizontality.

Thus, when we see such strata inclined instead of horizontal, evidence of an episode of deformation. In order to uniquely define the orientation of a planar feature we first need to define two terms - strike and dip.

For an inclined plane the *strike* is the compass direction of any horizontal line on the plane. The *dip* is the angle between a horizontal plane and the inclined plane, measured perpendicular to the direction of strike.

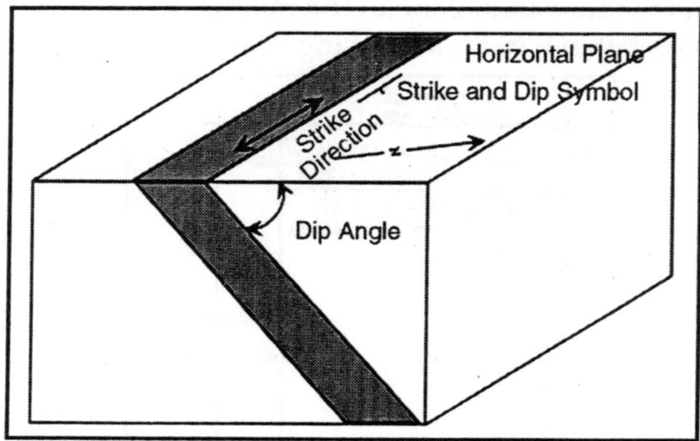

In recording strike and dip measurements on a geologic map, a symbol is used that has a long line oriented parallel to the compass direction of the strike.

A short tick mark is placed in the centre of the line on the side to which the inclined plane dips, and the angle of dip is recorded next to the strike and dip symbol as shown above. For beds with a 90^0 dip (vertical) the short line crosses the strike line, and for beds with no dip (horizontal) a circle with a cross inside is used as shown below.

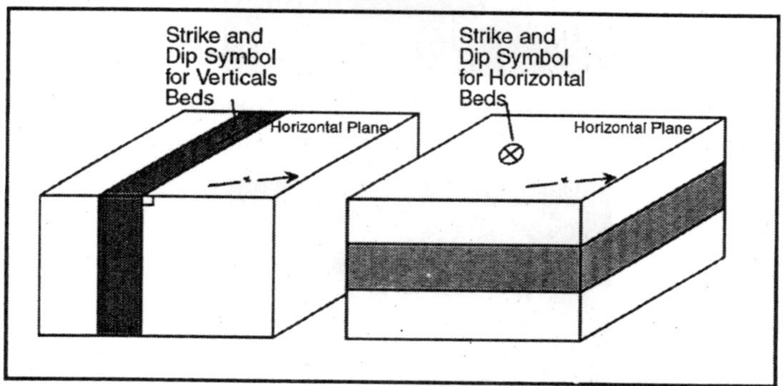

FAULTS

FEATURES

Rocks are very slowly, but continuously moving and changing shape. Under high temperature and pressure conditions common deep within Earth, rocks can bend and flow. In the cooler parts of Earth, rocks are colder and brittle and respond to large stresses by fracturing. Earthquakes are the agents of brittle rock failure.

A fault is a crack across which the rocks have been offset. They range in size from micrometres to thousands of kilometres in length and tens of kilometres in depth, but they are generally much thinner than they are long or deep. In addition to variation in size and orientation, different faults can accommodate different styles of rock deformation, such as compression and extension.

Not all faults intersect Earth's surface, and most earthquakes do no rupture the surface. When a fault does intersect the surface, objects may be offset or the ground may cracked, or raised, or lowered. We call a rupture of the surface by a fault a fault scarp and identifying scarps is an important task for assessing the seismic hazards in any region.

Fig. Fence Offset about 11 Feet During the 1906 San Francisco California Earthquake (Photo from the U.S. Geological Survey)

Fig. Fault Scarp.

EARTHQUAKES AND FAULTS

When an earthquake occurs only a part of a fault is involved in the rupture. That area is usually outlined by the distribution of aftershocks in the sequence.

We call the "point" (or region) where an earthquake rupture initiates the hypocentre or focus. The point on Earth's surface directly above the hypocentre is called the epicentre. When we plot earthquake locations on a map, we usually centre the symbol representing an event at the epicentre.

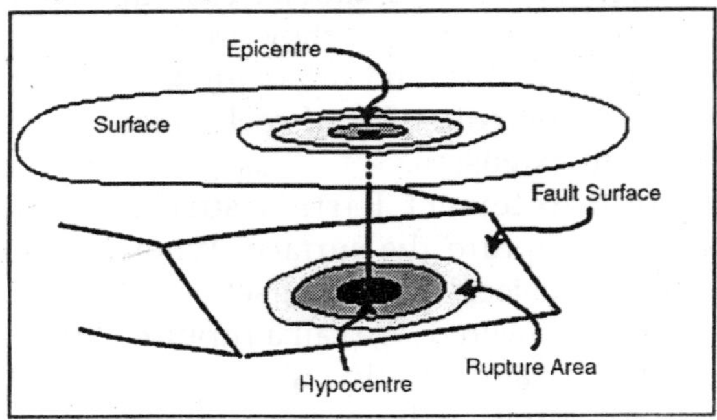

Generally, the area of the fault that ruptures increases with magnitude. Some estimates of rupture area are presented in the table below (The original data are from Wells and Coppersmith, Bulletin of the Seismological Society of America.

Date	Location	Length (km)	Depth (km)	(Mw)
04/18/06	San Francisco, CA	432	12	7.90
07/21/52	Kern County, CA	64	19	7.38
12/16/54	Fairview Peak, NV	50	15	7.17
12/16/54	Dixie Peak, NV	42	14	6.94
06/28/66	Parkfield, CA	35	10	6.25
02/09/71	San Fernando Valley, CA	17	14	6.64
10/28/83	Borah Peak, ID	33	20	6.93
10/18/89	Loma Prieta, CA	40	16	6.92
06/28/92	Landers, CA	62	12	7.34

Although the exact area associated with a given size earthquake varies from place to place and event to event, we can make predictions for "typical" earthquakes based on the available observations.

Magnitude	Fault Dimensions (Length x Depth, in km)
4.0	1.2 × 1.2
5.0	3.3 × 3.3
6.0	10 × 10
6.5	16 × 16, 25 × 10
7.0	40 × 20, 50 × 15
7.5	140 × 15, 100 × 20, 72 × 30, 50 × 40, 45 × 45
8.0	300 × 20, 200 × 30, 150 × 40, 125 × 50

These numbers should give you a rough idea of the size of structure that we are talking about when we discuss earthquakes.

FAULT STRUCTURE

Although the number of observations of deep fault

structure is small, the available exposed faults provide some information on the deep structure of a fault. A fault "zone" consists of several smaller regions defined by the style and amount of deformation within them.

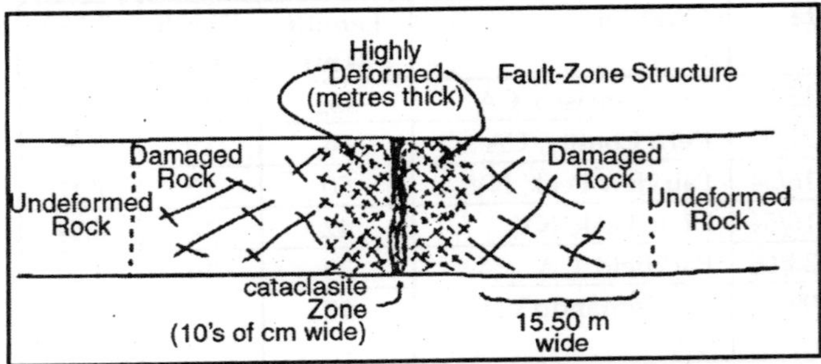

Fig. Structure of an Exposed Section

The centre of the fault is the most deformed and is where most of the offset or slip between the surrounding rock occurs. The region can be quite small, about as wide as a pencil is long, and it is identified by the finely ground rocks called cataclasite (we call the ground up material found closer to the surface, gouge). From all the slipping and grinding, the gouge is composed of very fine-grained material that resembles clay. Surrounding the central zone is a region several metres across that contains abundant fractures. Outside that region is another that contains distinguishable fractures, but much less dense than the preceding region. Last is the competent "host" rock that marks the end of the fault zone.

FAULT CLASSIFICATIONS

Active Faults

Active faults are structure along which we expect displacement to occur. By definition, since a shallow earthquake is a process that produces displacement across a fault, all shallow earthquakes occur on active faults.

Inactive Faults

Inactive faults are structures that we can identify, but which do no have earthquakes. As you can imagine, because of the complexity of earthquake activity, judging a fault to be inactive can be tricky, but often we can measure the last time substantial offset occurred across a fault.

If a fault has been inactive for millions of years, it's certainly safe to call it inactive. However, some faults only have large earthquakes once in thousands of years, and we need to evaluate carefully their hazard potential.

Reactivated Faults

Reactivated faults form when movement along formerly inactive faults can help to alleviate strain within the crust or upper mantle. Deformation in the New Madrid seismic zone in the central United States is a good example of fault reactivation. Structure formed about 500 Ma ago are responding to a new forces and relieving strain in the mid-continent.

FAULTING GEOMETRY

Faulting is a complex process and the variety of faults that exists is large. We will consider a simplified but general fault classification based on the geometry of faulting, which we describe by specifying three angular measurements: dip, strike, and slip.

Dip

The fault illustrated in the previous section was oriented vertically. In Earth, faults take on a range of orientations from vertical to horizontal. Dip is the angle that describes the steepness of the fault surface.

This angle is measured from Earth's surface, or a plane parallel to Earth's surface. The dip of a horizontal fault is zero (usually specified in degrees: 0°), and the dip of a vertical fault is 90°. We use some old mining terms to label the rock "blocks" above and below a fault. If you were tunneling through a fault, the material beneath the fault would be by your feet, the other material would be hanging above you head. The material resting on the fault is called the hanging wall, the material beneath the fault is called the foot wall.

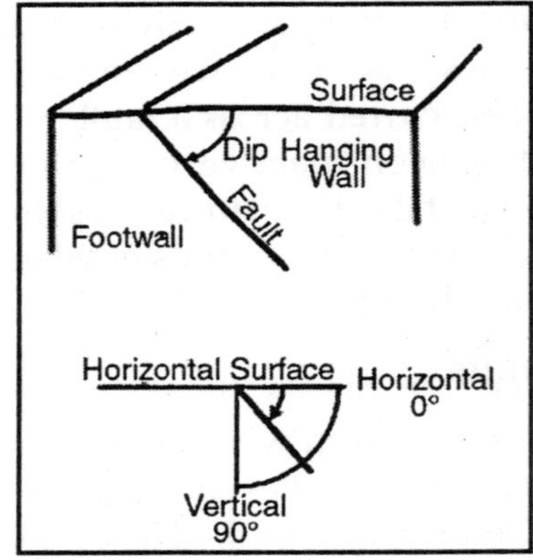

Strike

The strike is an angle used to specify the orientation of the fault and measured clockwise from north. For example, a strike of 0° or 180° indicates a fault that is oriented in a north-south direction, 90° or 270° indicates east-west oriented structure. To remove the ambiguity, we always specify the strike such that when you "look" in the strike direction, the fault dips to you right. Of course if the fault is perfectly vertical you have to describe the situation as a special case. If a fault curves, the strike varies along the fault, but this is seldom causes a communication problem if you are careful to specify the location of the measurement.

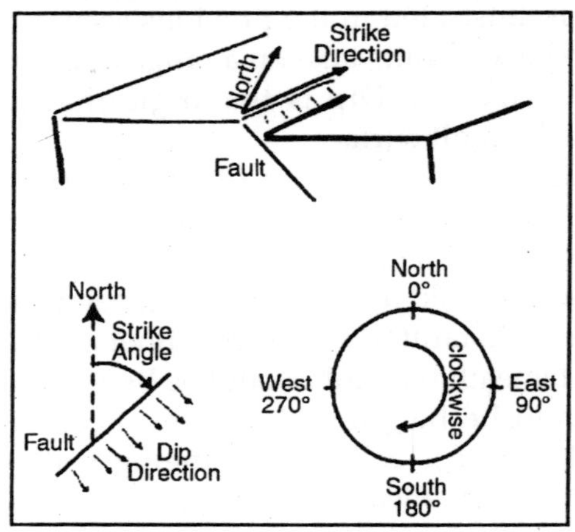

Slip

Dip and strike describe the orientation of the fault, we also have to describe the direction of motion across the fault. That is, which way did one side of the fault move with respect to the other. The parametre that describes this motion is called the slip. The slip has two components, a "magnitude" which tells us how far the rocks moved, and a direction (it's a vector). We usually specify the magnitude and direction separately.

The magnitude of slip is simply how far the two sides of the fault moved relative to one another; it's a distance usually a few centimetres for small earthquakes and metres for large events. The direction of slip is measured on the fault surface, and like the strike and dip, it is specified as an angle. Specifically the slip direction is the direction that the hanging wall moved relative to the footwall. If the hanging wall moves to the right, the slip direction is 0°; if it moves up, the slip angle is 90°, if it moves to the left, the slip angle is 180°, and if it moves down, the slip angle is 270° or -90°.

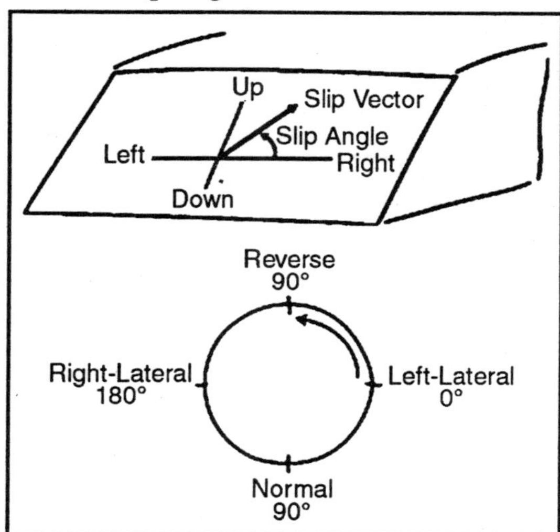

Hanging wall movement determines the geometric classification of faulting. We distinguish between "dip-slip" and "strike-slip" hanging-wall movements.

Dip-slip movement occurs when the hanging wall moved

predominantly up or down relative to the footwall. If the motion was down, the fault is called a normal fault, if the movement was up, the fault is called a reverse fault. Downward movement is "normal" because we normally would expect the hanging wall to slide downward along the foot wall because of the pull of gravity. Moving the hanging wall up an inclined fault requires work to overcome friction on the fault and the downward pull of gravity.

When the hanging wall moves horizontally, it's a strike-slip earthquake. If the hanging wall moves to the left, the earthquake is called right-lateral, if it moves to the right, it's called a left-lateral fault. The way to keep these terms straight is to imagine that you are standing on one side of the fault and an earthquake occurs. If objects on the other side of the fault move to your left, it's a left-lateral fault, if they move to your right, it's a right-lateral fault.

When the hanging wall motion is neither dominantly vertical nor horizontal, the motion is called oblique-slip. Although oblique faulting isn't unusual, it is less common than the normal, reverse, and strike-slip movement.

Fault Styles

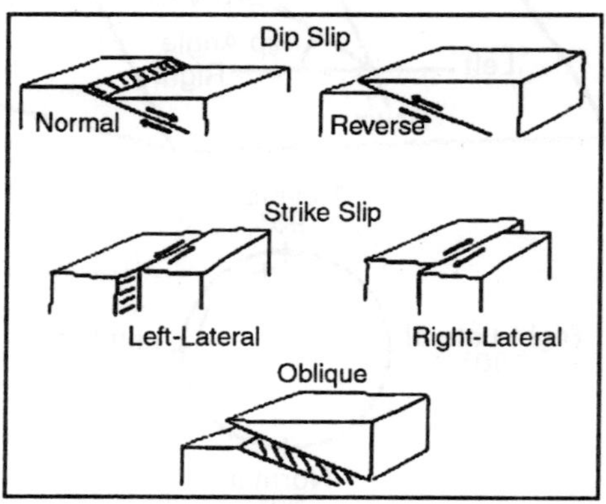

FAULTS AND FORCES

The style of faulting is an indicator of rock deformation and reflects the type of forces pushing or pulling on the region.

Near Earth's surface, the orientation of these forces are usually oriented such that one is vertical and the other two are horizontal. The precise direction of the horizontal forces varies from place to place as does the size of each force.

The style of faulting that is a reflection of the relative size of the different forces - in particular is the relative size of the vertical to the horizontal forces. There are three cases to consider, the vertical force can be the smallest, the largest, or the intermediate. If the vertical force is the largest, we get normal faulting, if it is the smallest, we get reverse faulting. When the vertical force is the intermediate force, we get strike-slip faulting. Normal faulting is indicative of a region that is stretching, and on the continents, normal faulting usually occurs in regions with relatively high elevation such as plateaus.

Reverse faulting reflects compressive forces squeezing a region and they are common in uplifting mountain ranges and along the coast of many regions bordering the Pacific Ocean. The largest earthquakes are generally low-angle (shallow dipping) reverse faults associated with "subduction" plate boundaries. Strike-slip faulting indicates neither extension nor compression, but identifies regions where rocks are sliding past each other. The San Andreas fault system is a famous example of strike-slip deformation - part of coastal California is sliding to the northwest relative to the rest of North America - Los Angeles is slowly moving towards San Francisco.

As you might expect, the distribution of faulting styles is not random, but varies systematically across Earth and was one of the most important observations in constructing the plate tectonic model which explains so much of what we observe happening in the shallow part of Earth.

Fault Type	Normal Faulting	Reverse Faulting	Transform Faulting
Deformation Style	Extension	Compression	Translation
Force	Vertical	Vertical	Vertical
Orientation	Force Is Largest	Force Is Smallest	Force Is Intermediate

EARTHQUAKE FOCAL MECHANISMS

We use a specific set of symbols to identify faulting geometry on maps. The symbols are called earthquake focal mechanisms or sometimes "seismic beach balls". A focal mechanism is a graphical summary the strike, dip, and slip directions.

An earthquake focal mechanism is a projection of the intersection of the fault surface and an imaginary lower hemisphere (we'll use the lower hemisphere, but we could also use the upper hemisphere), surrounding the centre of the rupture.

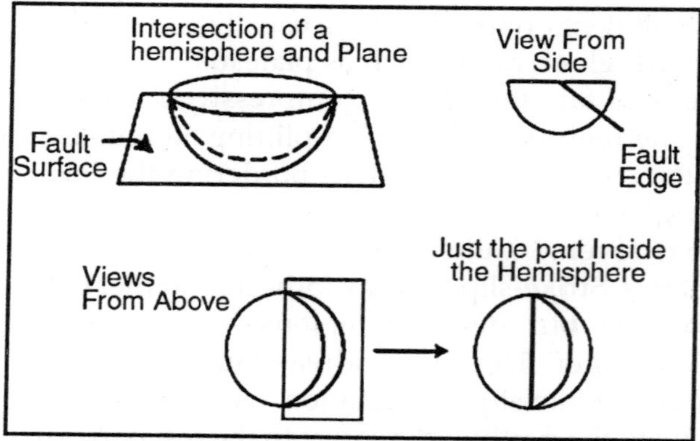

The intersection between the fault "plane" and the sphere is a curve. The focal mechanism shows the view of the hemisphere from directly above. We can show the orientation of a plane (i.e. the strike and dip) using just one curve, to include information on the slip, we use two planes and shade opposite quadrants of the hemisphere.

The price we pay for the ability to represent slip is that you cannot identify which of the two planes on the focal mechanism is the fault without additional information (such as the location and trend of aftershocks).

Some example focal mechanisms are shown below: You should memorize the top three, which correspond to dip-slip reverse and normal faulting on a fault dipping 45°, and strike-slip faulting on a vertical fault. The lower two mechanisms correspond to a low-angle reverse earthquake (the dip is low) and the last example is an oblique event with components of both strike-slip and dip-slip movement.

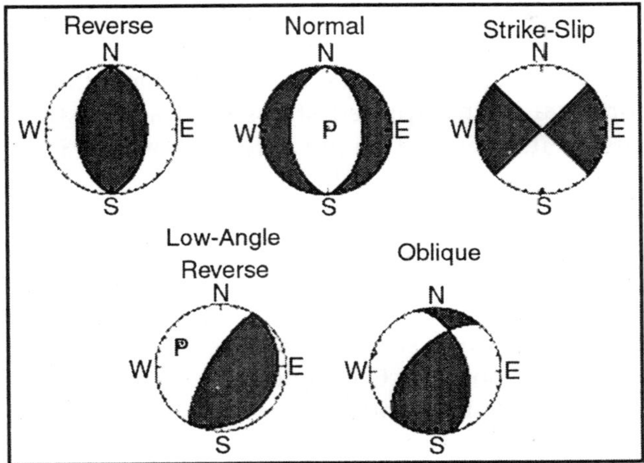

The strike of any plane can be read from a focal mechanism by identifying the intersection of the fault (shown as the boundary between shaded and unshaded regions) with the circle surrounding the mechanism (and using the dip-to-the-right rule).

STRESS AND STRAIN

Stress

Stress is a force per unit area or a force that acts on a surface. When I described the types of forces associated with the different styles of faulting (in the section "Faults and Faulting"), I was describing stresses (the force per unit area on the fault).

Friction

Friction is a stress which resists motion and acts in all natural systems.

For earthquake studies, friction on faults and the orientation and relative magnitudes of the "regional" stresses that determine the style of faulting are of primary interest and importance.

Strain

Is a measure of material deformation such as the amount of compression when you squeeze or the amount of elongation when you stretch something.

In elastic deformation the amount of elongation is linearly proportional to the applied stress, and an elastic material returns to its original shape after the stress is relieved. Additionally, a strained, elastic material stores the energy used to deform it, and that energy is recoverable.

ELASTIC REBOUND

As you know, some regions repeatedly experience earthquakes and this suggests that perhaps earthquakes are part of a cycle. The effects of repeated earthquakes were first noted late in the nineteenth century by American geologist G. K. Gilbert. Gilbert observed a fresh fault scarp following the 1872 Owens Valley, California earthquake and correlated the scarp and uplift from a single earthquake with the uplift of the Sierra Nevada mountains.

Decades later, following the 1906 San Francisco, California earthquake, H. F. Reid presented a similar hypothesis to explain better-documented movements along coastal California measured both before and after a large earthquake. Reid's model of the earthquake cycle has become known as the "Elastic Rebound Model".

The key to Reid's success was the availability of "before" and "after" observations for the earthquake which allowed him to see strain build in the crust before the event, and then see that strain release during the earthquake.

In the diagram, we have two blocks of rock separated by a fault. As the two blocks move in opposite directions, friction acting on the fault resists movement and keeps the two sides from sliding. The rock strains as elastic energy is added, eventually, the strain loads the fault too much and overcomes the frictional "strength" of the fault. The rocks on either side of the fault jerk past each other in an earthquake. The earthquake releases the stored elastic strain energy as heat along the fault and as seismic vibrations.

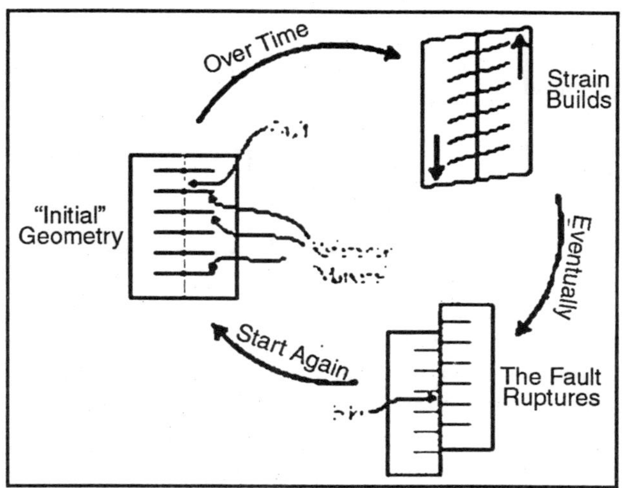

For an ideal elastic-rebound fault, the stress on the fault periodically cycles between a minimum and maximum value and if the two blocks continue to move at a constant rate, the recurrence time (the time between earthquakes) is also uniform.

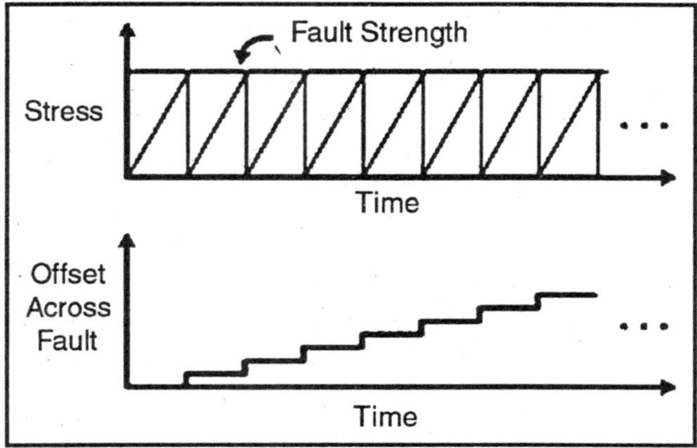

Unfortunately, actual faults are more complex, and the recurrence time is not periodic (which is one reason why earthquake prediction is so difficult). We have few observations of complete earthquake cycles because earthquakes take so long to recur.

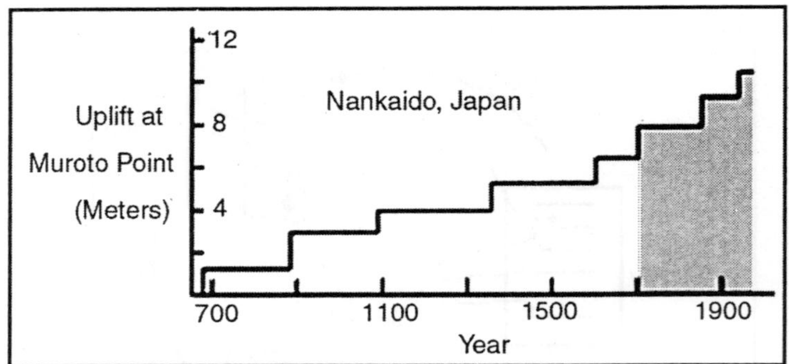

The Figure above shows the observations from the Nankaido region of Japan (the gray region, the older values are estimated from earthquake histories), one of the few regions where observations on strain throughout several earthquake cycles exist. You can see that neither the time nor the slip is uniform from earthquake-to-earthquake.

Faults occur when brittle rocks fracture and there is an offset along the fracture. When the offset is small, the displacement can be easily measured, but sometimes the displacement is so large that it is difficult to measure.

TYPES OF FAULTS

Faults can be divided into several different types depending on the direction of relative displacement. Since faults are planar features, the concept of strike and dip also applies, and thus the strike and dip of a fault plane can be measured.

One division of faults is between dip-slip faults, where the displacement is measured along the dip direction of the fault, and strike-slip faults where the displacement is horizontal, parallel to the strike of the fault.

Dip Slip Faults

Dip slip faults are faults that have an inclined fault plane and along which the relative displacement or offset has occurred along the dip direction.

Note that in looking at the displacement on any fault we don't know which side actually moved or if both sides moved, all we can determine is the relative sense of motion. For any inclined fault plane we define the block above the fault as the *hanging wall block* and the block below the fault as the *footwall block.*

Normal Faults

Are faults that result from horizontal tensional stresses in brittle rocks and where the hanging-wall block has moved down relative to the footwall block.

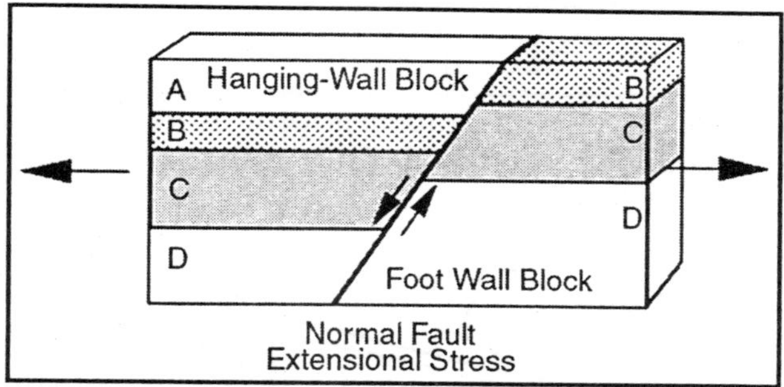

Normal Fault
Extensional Stress

Horsts and Gabens

Due to the tensional stress responsible for normal faults, they often occur in a series, with adjacent faults dipping in opposite directions.

In such a case the down-dropped blocks form *grabens* and the uplifted blocks form *horsts.* In areas where tensional stress has recently affected the crust, the grabens may form *rift valleys*and the uplifted horst blocks may form linear mountain ranges. The East African Rift Valley is an example of an area where continental extension has created such a rift. The basin and range province of the western U.S. (Nevada, Utah, and Idaho) is also an area that has recently undergone crustal extension. In the basin and range, the basins are elongated grabens that now form valleys, and the ranges are uplifted horst blocks.

A normal fault that has a curved fault plane with the dip decreasing with depth can cause the down-dropped block to rotate.

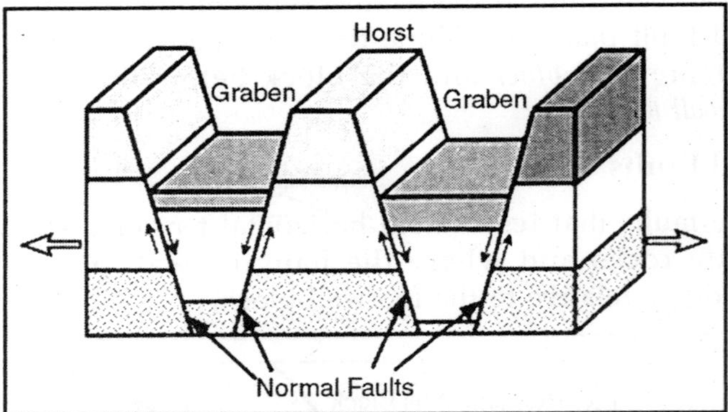

In such a case a half-graben is produced, called such because it is bounded by only one fault instead of the two that form a normal graben.

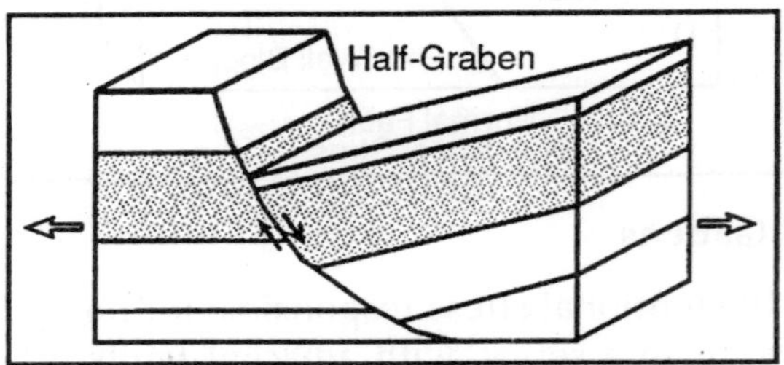

Reverse Faults

Are faults that result from horizontal compressional stresses in brittle rocks, where the hanging-wall block has moved up relative the footwall block.

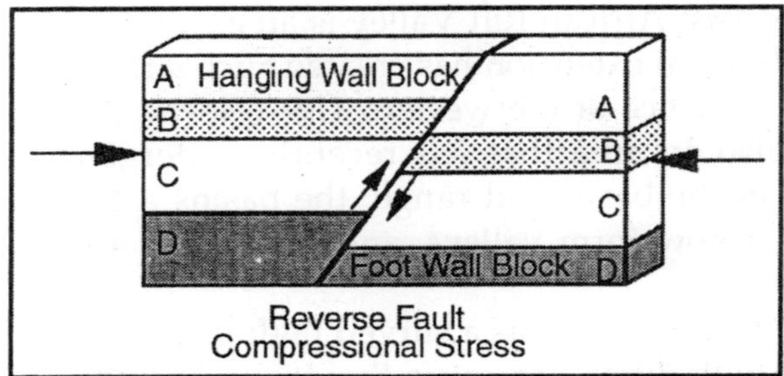

A *Thrust Fault* is a special case of a reverse fault where the dip of the fault is less than 15°. Thrust faults can have considerable displacement, measuring hundreds of kilometres, and can result in older strata overlying younger strata.

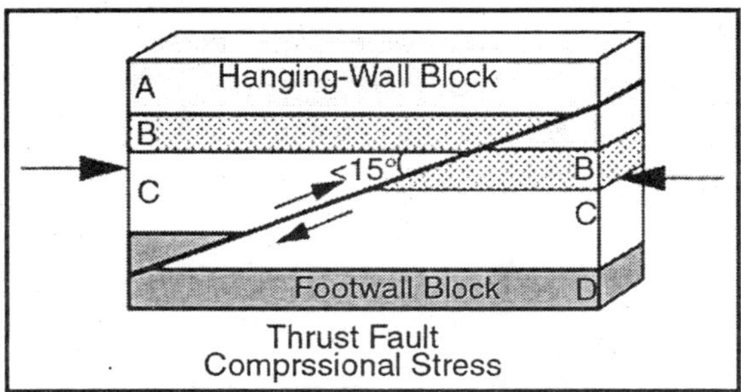

Thrust Fault
Comprssional Stress

Strike Slip Faults

Are faults where the relative motion on the fault has taken place along a horizontal direction. Such faults result from shear stresses acting in the crust. Strike slip faults can be of two varieties, depending on the sense of displacement. To an observer standing on one side of the fault and looking across the fault, if the block on the other side has moved to the left, we say that the fault is a *left-lateral strike-slip fault*.

If the block on the other side has moved to the right, we say that the fault is a *right-lateral strike-slip fault*. The famous San Andreas Fault in California is an example of a right-lateral strike-slip fault. Displacements on the San Andreas fault are estimated at over 600 km.

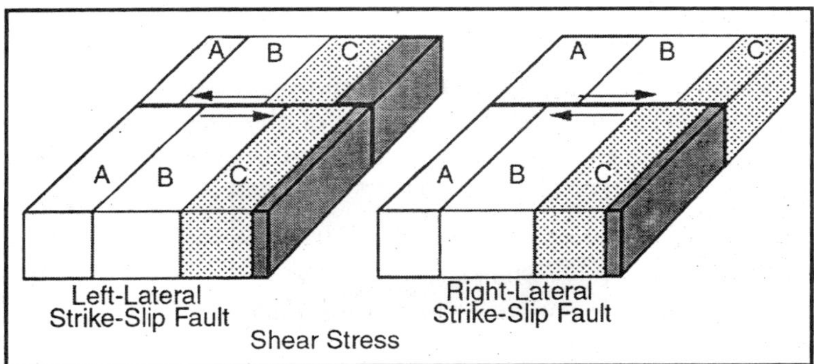

Left-Lateral Strike-Slip Fault
Right-Lateral Strike-Slip Fault
Shear Stress

Transform-Faults are a special class of strike-slip faults. These are plate boundaries along which two plates slide past one another in a horizontal manner. The most common type of transform faults occur where oceanic ridges are offset. Note that the transform fault only occurs between the two segments of the ridge. Outside of this area there is no relative movement because blocks are moving in the same direction. These areas are called fracture zones. The San Andreas fault in California is also a transform fault.

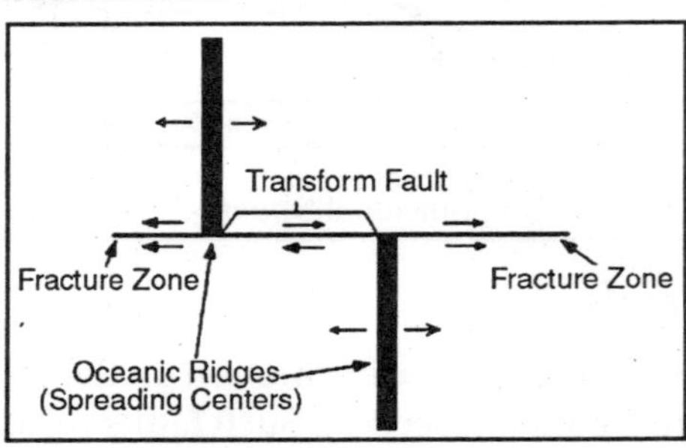

EVIDENCE OF MOVEMENT ON FAULTS

- *Slikensides* are scratch marks that are left on the fault plane as one block moves relative to the other. Slickensides can be used to determine the direction and sense of motion on a fault.
- *Fault Breccias* are crumbled up rocks consisting of angular fragments that were formed as a result of grinding and crushing movement along a fault.

FOLD

INTRODUCTION

Folds are amongst the most common tectonic structures found in rocks, and can make some of the most spectacular features. They can occur on all scales and in all environments.

In areas of active mountain building, as for example here in the front ranges of the Himalayas in Pakistan (right), folds bulge up the landscape. The upward moving parts (called antiforms) create hills while the areas that have gone down relative to the hills (called synforms) collect detritus eroded from the surrounding countryside.

A common misconception is that rocks can only fold when they are nearly molten. These outcrops tell a different story. Rocks like these limestone's can fold even at the Earth's surface - provided they are given enough time. A fold like this could take a hundred thousand years to grow. But if the rocks were pushed together more quickly - they could break - makingfaults.

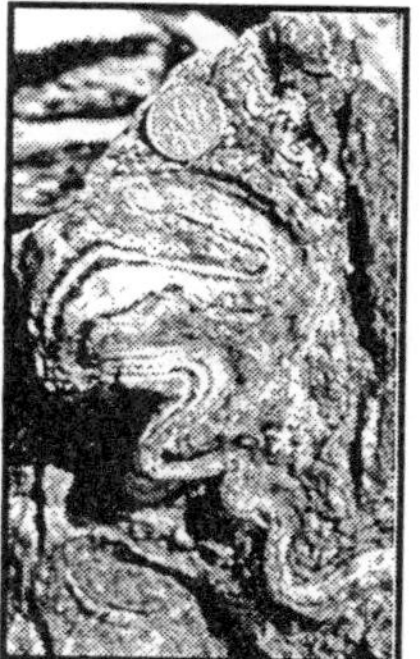

Folds can occur on all scales. These folds from the island of Syros only have a wavelength of a few millimetres. To make folds the rocks must have a mechanical layering. If they are homogeneous they will just squash together without folding. A stiff layer (more viscous) will fold if embedded in weaker (less viscous) material. You can try this yourself - using different types of plasticene.

DESCRIVING FOLDS

Folds are wonderful things and highly variable. There's a plethora of jargon for describing them. More usefully, there are many ways of measuring them so that we can quantify their shape and orientation. From these careful descriptions we can learn more about how folds form and what they might be telling us about the larger-scale tectonics.

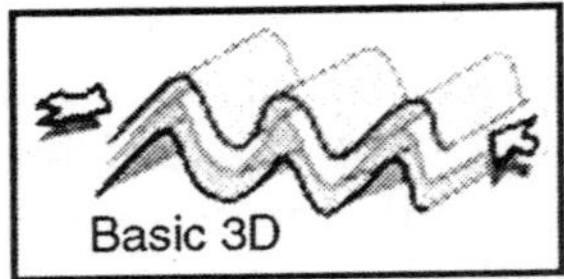

Basic geometry of folds as wave-forms. Covers hinges and axial surfaces too.

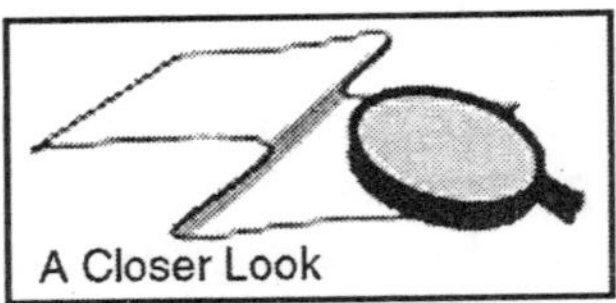

How to classify folds - going into more detail.

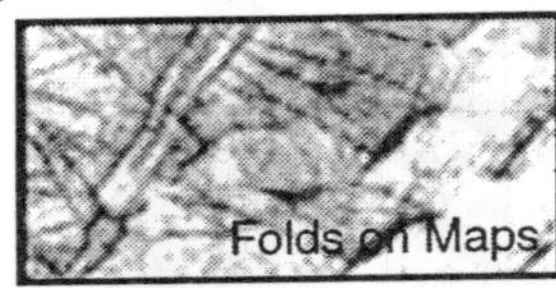

Quantitative approaches to deducing fold geometries from map-patterns, using structure contours. A real 3D insight!

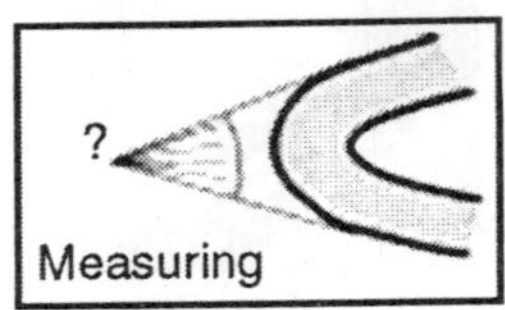

What to measure on folds - orientations, wavelength and amplitude.

How to use small-scale observations of folded rocks to gain insight on the larger-scale structure.

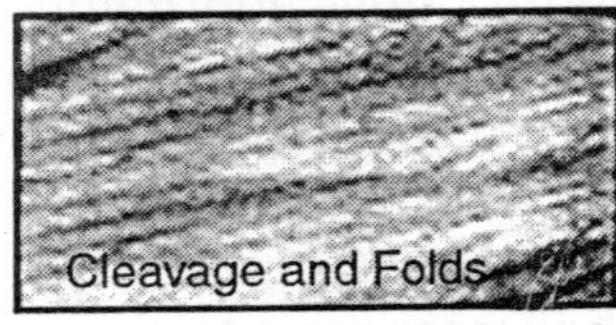

How cleavage can be used to deduce larger-scale structur

FOLDING OF DUCTILE ROCKS

When rocks deform in a ductile manner, instead of fracturing to form faults, they may bend or fold, and the resulting structures are called *folds*. Folds result from compressional stresses acting over considerable time. Because the strain rate is low, rocks that we normally consider brittle can behave in a ductile manner resulting in such folds. We recognize several different kinds of folds.

Monoclines

They are the simplest types of folds. Monoclines occur when horizontal strata are bent upward so that the two limbs of the fold are still horizontal.

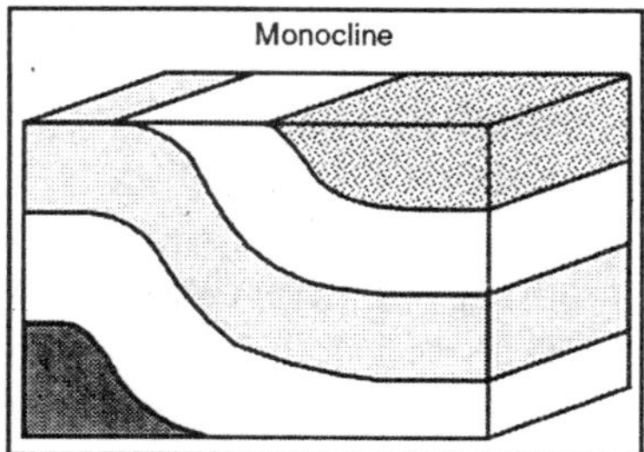

Anticlines are folds where the originally horizontal strata has been folded upward, and the two limbs of the fold dip away from the hinge of the fold.

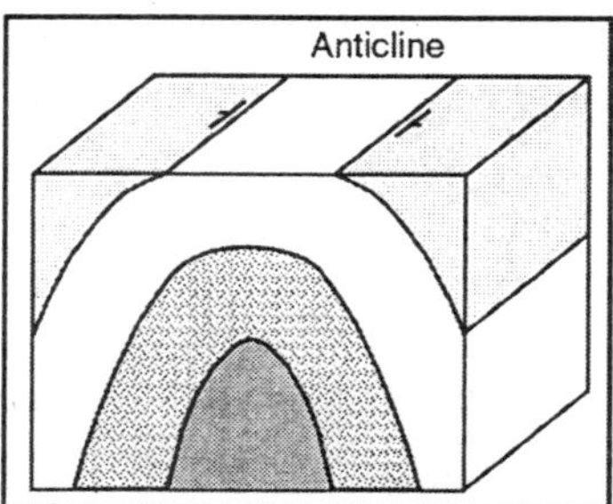

Synclines

Are folds where the originally horizontal strata have been folded downward, and the two limbs of the fold dip inward toward the hinge of the fold. Synclines and anticlines usually occur together such that the limb of a syncline is also the limb of an anticline.

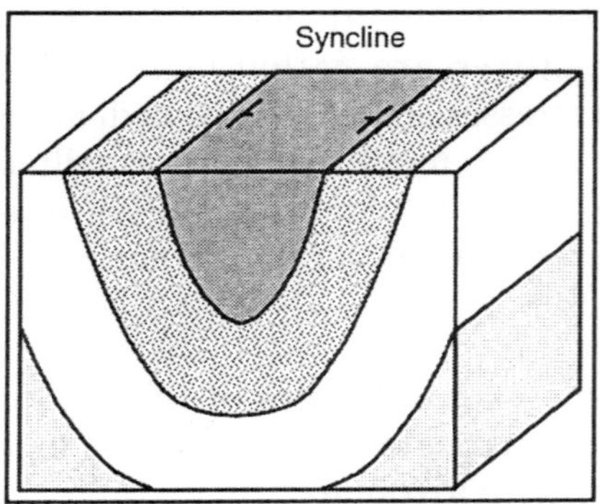

Geometry of Folds

Folds are described by their form and orientation. The sides of a fold are called *limbs*. The limbs intersect at the tightest part of the fold, called the *hinge*. A line connecting all points on the hinge is called the *fold axis*. In the diagrams above, the fold axes are horizontal, but if the fold axis is not horizontal the fold is called a *plunging fold* and the angle that the fold axis makes with a horizontal line is called the *plunge* of the fold. An imaginary plane that includes the fold axis and divides the fold as symmetrically as possible is called the *axial plane* of the fold.

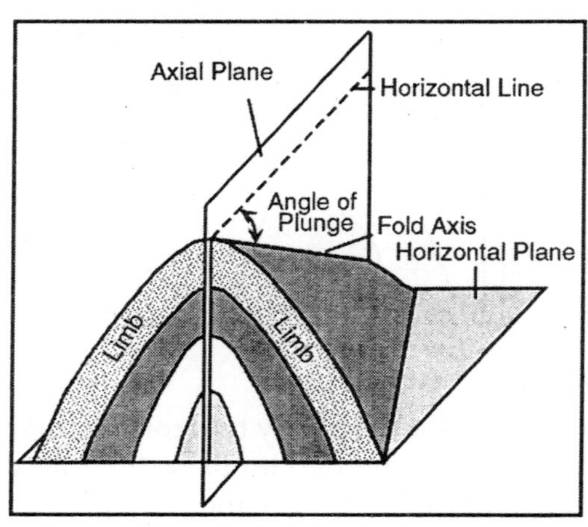

Note that if a plunging fold intersects a horizontal surface, we will see the pattern of the fold on the surface.

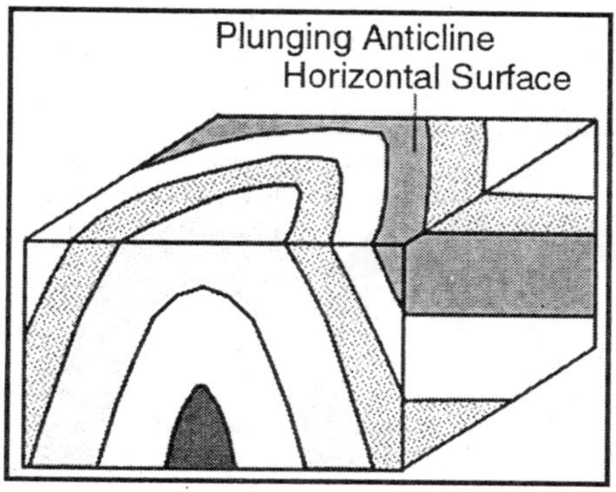

Classification of Folds

Folds can be classified based on their appearance:

- If the two limbs of the fold dip away from the axis with the same angle, the fold is said to be a *symmetrical fold*.
- If the limbs dip at different angles, the folds are said to be *asymmetrical folds*.
- If the compressional stresses that cause the folding are intense, the fold can close up and have limbs that are parallel to each other. Such a fold is called an *isoclinal fold* (iso means same, and cline means angle, so isoclinal means the limbs have the same angle). Note the isoclinal fold depicted in the diagram below is also a symmetrical fold.
- If the folding is so intense that the strata on one limb of the fold becomes nearly
- upside down, the fold is called an *overturned fold*.
- An overturned fold with an axial plane that is nearly horizontal is called a *recumbant fold*.
- A fold that has no curvature in its hinge and straight-sided limbs that form a zigzag pattern is called a *chevron fold*.

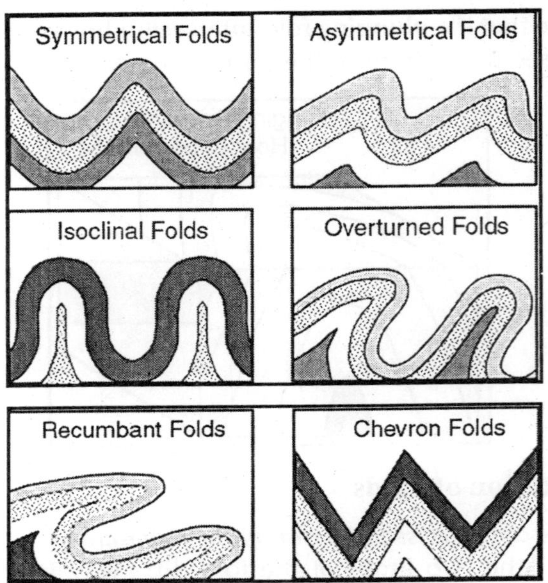

Relationship Between Folding and Faulting

Because different rocks behave differently under stress, we expect that some rocks when subjected to the same stress will fracture or fault, while others will fold.

When such contrasting rocks occur in the same area, such as ductile rocks overlying brittle rocks, the brittle rocks may fault and the ductile rocks may bend or fold over the fault.

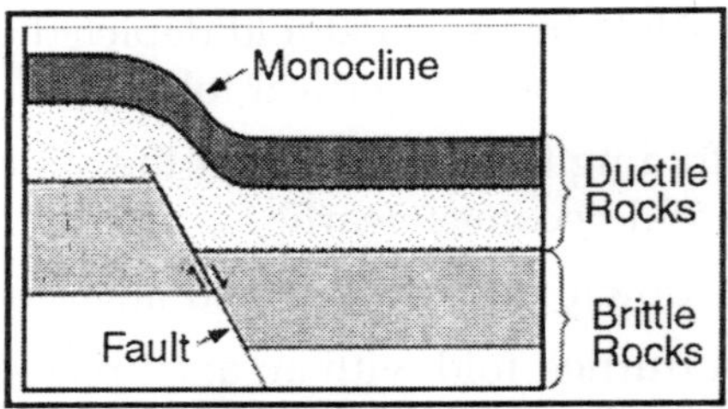

Also since even ductile rocks can eventually fracture under high stress, rocks may fold up to a certain point then fracture to form a fault.

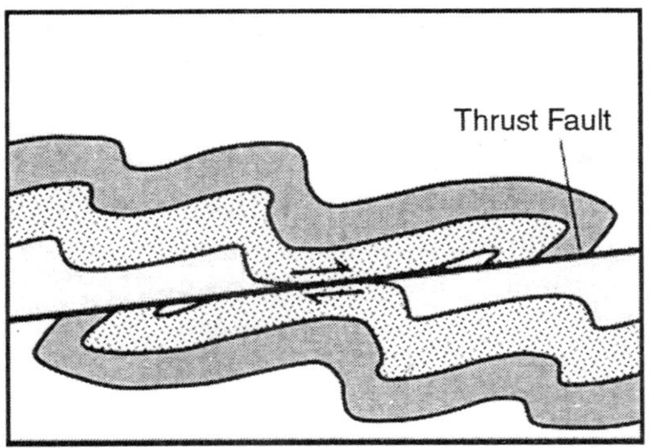

FOLDS AND TOPOGRAPHY

Since different rocks have different resistance to erosion and weathering, erosion of folded areas can lead to a topography that reflects the folding. Resistant strata would form ridges that have the same form as the folds, while less resistant strata will form valleys (see figure).

MOUNTAIN RANGES

One of the most spectacular results of deformation acting within the crust of the Earth is the formation of mountain ranges. Mountains originate by three processes, two of which are directly related to deformation. Thus, there are three types of mountains

Fault Block Mountains

As the name implies, fault block mountains originate by faulting. As discussed previously, both normal and reverse faults can cause the uplift of blocks of crustal rocks. The Sierra Nevada mountains of California, and the mountains in the Basin and Range province of the western U.S., as discussed previously, were formed by faulting processes and are thus fault block mountains.

Fold and Thrust Mountains

Large compressional stresses can be generated in the crust by tectonic forces that cause continental crustal areas to collide.

When this occurs the rocks between the two continental blocks become folded and faulted under compressional stresses and are pushed upward to form fold and thrust mountains. The Himalayan Mountains (currently the highest on Earth) are mountains of this type and were formed as a result of the Indian Plate colliding with the Eurasian plate. Similarly the Appalachian Mountains of North America and the Alps of Europe were formed by such processes.

Volcanic Mountains

The third type of mountains, volcanic mountains, are not formed by deformational processes, but instead by the outpouring of magma onto the surface of the Earth. The Cascade Mountains of the western U.S., and of course the mountains of the Hawaiian Islands and Iceland are volcanic mountains.

JOINTS UNCONFORMITY AND CLASSIFICATION

JOINTS

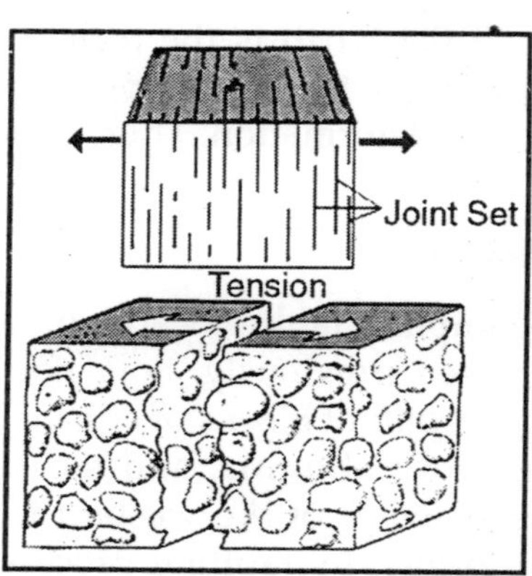

joint, in geology, fracture in rocks along which no appreciable movement has occurred. Nearly vertical, or sheet, joints that result from shrinkage during cooling are commonly found in igneous rocks. Similar joints occur in thick beds of sandstone and gneiss, with the sheets resembling the structure of a sliced onion. The prismatic joints of the Palisades of New

Jersey and Devil's Tower, Wyoming, are examples of joints caused by contraction during the cooling of fine-grained igneous rock masses. Deep-seated igneous rocks often have joints approximately parallel to the surface, suggesting that they formed by expansion of the rock mass as overlying rocks were eroded away.

Some joints in sedimentary rocks may have formed as the result of contraction during compaction and drying of the sediment. In some cases, jointing of the rock may result from the action of the same forces that cause folds and faults. In relatively undisturbed sedimentary rocks, such joints are often in two vertical sets perpendicular to one another. Commonly, streams develop along zones of weakness caused by joints in rocks, and thus the regional pattern of joint orientation often exerts a strong control on the development of drainage patterns.

Cause Of Joints (Rocks)

With regard to the manner of their production, joints may be classified into two series: those which are due to tension, the rock usually parting in planes normal to the directions of tension; those which are due to compression, the cracks forming in the shearing-planes.

Tension Joints

In igneous rocks joints are caused by the cooling and consequent contraction of the highly heated mass. This shrinkage sets up tensile stresses in the mass to which the rock yields by cracking and parting, the shape of the blocks being largely controlled by the coarseness or fineness of the mass. Igneous rocks are subject to all the vicissitudes which affect other kinds of rocks; they are faulted, compressed, exposed to tension, etc. Hence, systems of joints may occur in them, which were formed subsequently to the shrinkage-joints due to the contraction of cooling. In some cases the jointing ofsedimentary rocks may perhaps be caused by a shrinkage of the mass on drying, but this cannot be an important method of producing systems of joints.

The convex sides of anticlinal and synclinal folds are stretched, and (provided they are not too deeply buried) the stretching may result in a system of cracks radial to the curves which follow the strike of the beds. Folds are not horizontal, but pitch in the direction of their axes. This complex folding may produce two sets of tensile stresses perpendicular to each other, and thus cause two series of joints, one following the strike and the other the dip of the beds.

Complex folding must produce a twisting and warping of the strata, and it has been experimentally shown that a brittle substance, when twisted, cracks in two sets of fractures which intersect nearly at right angles. How slight is the twisting and warping needful to produce joints is shown by the fact that strata which are perfectly horizontal, so far as can be detected, are jointed.

The modern limestones which are formed in coral-reefs are jointed, even in cases where the movements resulting in fracture must have been minimal. Tension joints produce either rough, or smooth and sharply cut surfaces, which is determined by the character of the rock. In sandstones which are weakly cemented the cracks pass between the grains, while in hard and firm rocks the fractures are clean.

Fig. Jointing in Limestone, Black Hills,

Compression Joints

Compression Joints are caused when the rocks yield along the shearing-planes. In simply folded strata are produced two sets of strike joints which are inclined toward each other, but whether dip joints will be made by complex folding is not certain. In some conglomerates the joint planes pass through the hard quartz pebbles and leave a smooth, even, shining face. Tension would pull such a pebble out of its socket and only by shearing could it be cleanly cut.

The whole subject of joints in sedimentary rocks is a difficult one and the explanations given of them are not altogether satisfactory, for several other agencies may be involved in their production. It is, however, highly probable that the master joints which roughly follow the strike and dip of the strata, have been caused by the forces which produce folding.

Fig. Joints Dying away Downward,

Joints cannot occur in the shell of flowage, and are best developed in the shell of fracture, being of less importance in the transition belt between the two.

UNCONFORMITY

Relative and Absolute Age

In order to understand how geologists deal with time we first need to understand the concepts of relative age and absolute age.

Relative Age

Relative means that we can determine if something is younger than or older than something else. Relative time does not tell how old something is, all we know is the sequence of events. For example: the sandstone in this area is older than the limestone.

Absolute Age

Absolute age means that we can more or less precisely assign a number (in years, minutes, seconds, or some other units of time) to the amount of time that has passed. Thus we can say how old something is. For example: The sandstone is 300 million years old. To better understand these concepts, let's look at an archeological example: Imagine we are a group of archeologists studying two different trash pits recently discovered.

By carefully digging, we have found that each trash pit shows a sequence of layers. Although the types of trash in each pit is quite variable, each layer has a distinctive kind of trash that distinguishes it from other layers in the pits.

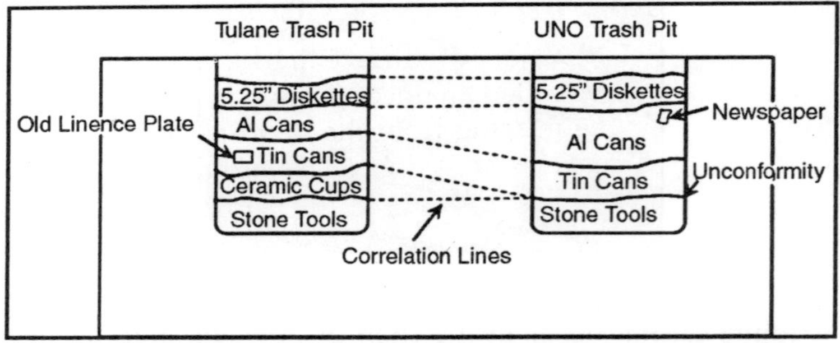

What can we say and learn from these excavations?

- *Relative age of trash layers*: Because of the shape of the pits the oldest layers of trash occur below younger layers i.e. the inhabitants of the area likely deposited the trash by throwing it in from the top, eventually filling the pits. Thus the relative age of the trash layers is, in order from youngest to oldest.:
- *5.25" Disk Layer*: Youngest
- Al Cans Layer
- Tin Cans Layer
- Ceramic Cups Layer
- *Stone Tools Layer*: Oldest

Notice that at this point we do not know exactly how old any layer really is.

Thus we do not know the absolute age of any given layer.

The civilizations that deposited the trash had a culture and industrial capabilities that evolved through time. The oldest inhabitants used primitive stone tools, later inhabitants used cups made of ceramics, even later inhabitants eventually used tin cans and then changed to Aluminum cans, and then they developed a technology that used computers.

Similar cultures must have existed in both areas and lived at the same time. Thus we can make correlation's between the layers found at the different sites, by reasoning that layers containing similar discarded items (artifacts) were deposited during the same time period.

Because the Ceramic Cups layer is found at the Tulane site, but not at the UNO site, the civilization that produced the Ceramic cups probably did not live in the UNO area. Thus, we can recognize a hiatus, or break in the depositional sequence at the UNO site. The surface marking in the break in deposition would be called an unconformity in geologic terms, and represents time missing from the depositional record.

The trash pits contain some clues to absolute age: The Tulane trash pit has an old license plate in the Tin Cans layer. This plate shows a date of 1950, thus the Tin Cans layer is about 48 years old. The UNO trash pit has an old newspaper in the Al Cans layer. The date on the newspaper is Oct. 1, 1978. Thus the Al Cans layer is about 20 years old.

In geology, we use similar principles to determine relative ages, correlations, and absolute ages.

- *Relative ages*: Principles of Stratigraphy
- *Correlations*: Fossils, key beds, physical criteria
- *Absolute ages*: Radiometric dating

Principles of Stratigraphy

Stratigraphy = the study of strata (layers) in the Earth's crust.

Laws of Stratigraphy

Original Horizontality: sedimentary strata are deposited in layers that are horizontal or nearly horizontal, parallel to or nearly parallel to the Earth's surface. Thus rocks that we now see inclined or folded have been disturbed since their original deposition.

Stratigraphic Superposition: Because of Earth's gravity, deposition of sediment will occur depositing older layers first followed by successively younger layers.

Thus, in a sequence of layers that have not been overturned by a later deformational event, the oldest layers will be on the bottom. This is the same principle used to determine relative age in the trash pits discussed previously. In fact, sedimentary rocks are, in a sense, trash from the Earth's surface deposited in basins. Breaks in the Stratigraphic Record. Because the Earth's crust is continually changing, i.e due to uplift, subsidence, and deformation, erosion is acting in some places and deposition of sediment is occurring in other places.

When sediment is not being deposited, or when erosion is removing previously deposited sediment, there will not be a continuous record of sedimentation preserved in the rocks. We call such a break in the stratigraphic record a hiatus (a hiatus was identified in our trash pit example by the non-occurrence of the Ceramic Cups layer at the UNO site). When we find evidence of a hiatus in the stratigraphic record we call it an unconformity. An unconformity is a surface of erosion or non-deposition. Three types of unconformities are recognized.

Angular Unconformity

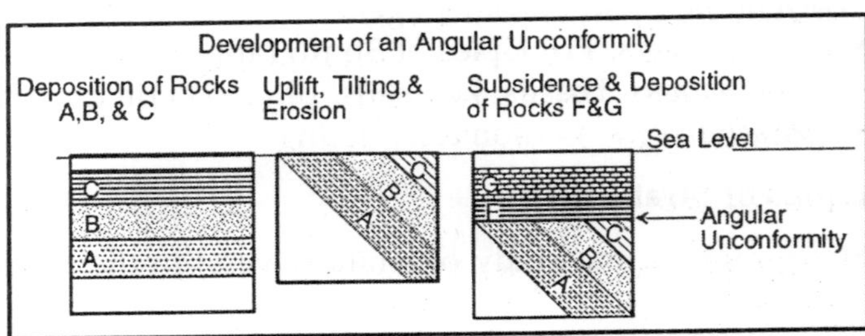

Because of the Laws of Stratigraphy, if we see a cross section like this in a road cut or canyon wall where we can recognize an angular unconformity, then we know the geologic history or sequence of events that must have occurred in the area to produce the angular unconformity. Angular unconformities are easy to recognize in the field because of the angular relationship of layers that were originally deposited horizontally

Disconformity

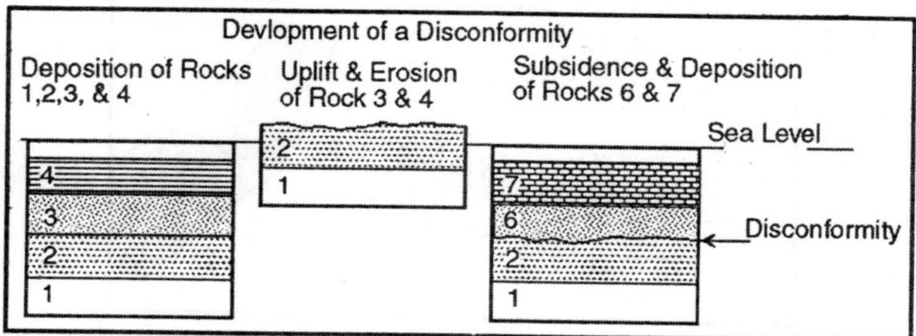

Disconformities (called parallel unconformities in your lab book) are much harder to recognize in the field, because often there is no angular relationship between sets of layers. Disconformities are usually recognized by correlating from one area to another and finding that some strata is missing in one of the areas. The unconformity recognized in the UNO trash pit is a disconformity.

Nonconformity

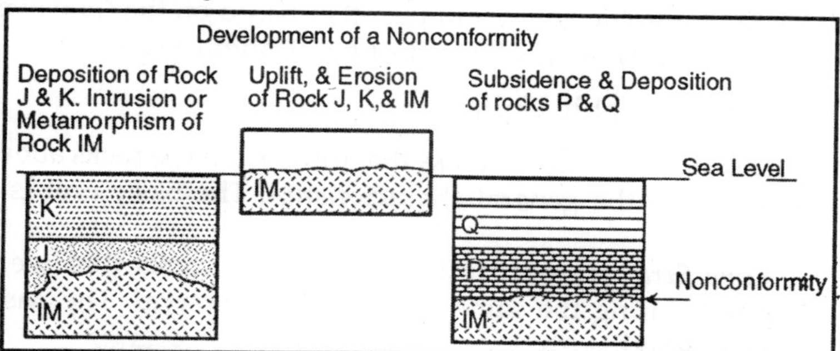

Nonconformities occur where rocks that formed deep in the Earth, such as intrusive igneous rocks or metamorphic rocks, are overlain by sedimentary rocks formed at the Earth's surface. The nonconformity can only occur if all of the rocks overlying the metamorphic or intrusive igneous rocks have been removed by erosion.

Variation of Unconformities:

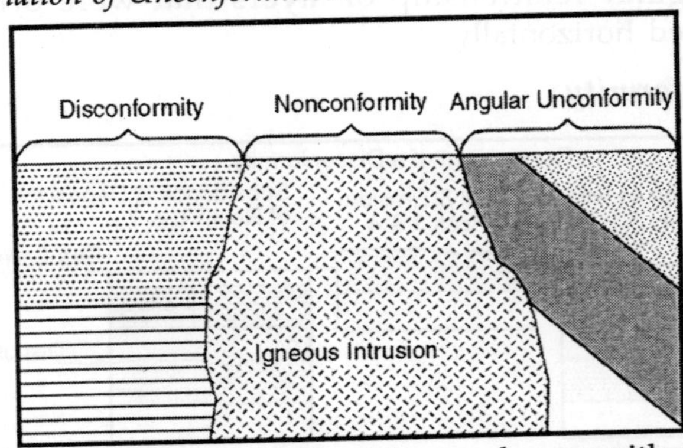

The nature of an unconformity can change with distance. Notice how if we are only examining a small area in the figure above, we would determine a different type of unconformity at each location, yet the unconformity itself was caused by the same erosional event.

CLASSIFICATION

Two types of stratigraphic classification are used, one based on physical characteristics or material properties of the rocks - Rock Stratigraphic Units, and the other based on the time over which the material was formed - time stratigraphic units.

Rock Stratigraphic Units

Distinctive bodies of rocks that differ from the rocks above and below in the general characteristics. The basic unit is a formation.

Time Stratigraphic Units: A bodies of rocks that were deposited during the same geologic time interval. The basic unit is a period.

Correlation of Rock Units

In order for rock units to be correlated over wide areas, they must be determined to be equivalent.

Determination of equivalence is based on:

- *Relative Age*: if two rock units are equivalent they must have the same age relative to rocks that occur below and above. The laws of stratigraphy enable this determination.

- *Physical Criteria*:
- Similarity in rock type, but only if relative age is equivalent. Key beds, like widespread volcanic ash layers that are the same over wide areas are often used to establish equivalence.

Fossils present - Fossils are key indicators of relative age (life has evolved through time) and environments of deposition.

The Geologic Column

Geologic Time Scale

Eon	Era	Period	Epoch	Age(my)
Phanerozoic (Visible Life)	Cenozoic (Recent Life) (Age of Mammals)	Quaternary	Holocene	0.01
			Pleistocene	1.6
		Tertiary	Pliocene	5.3
			Miocene	23.7
			Oligocene	36.6
			Eocene	57.8
			Paleocene	66.4
	Mesozoic (Middle Life) (Age of Reptiles)	Cretaceous		144
		Jurassic		208
		Triassiac		245
	Paleozoic (Ancient Life)	Permian		286
		Pennsylvanian		320
		Mississippian		360
		Devonian		408
		Silurian		438
		Ordovician		505
		Cambrian		570
Proter-ozoic (Early Life)	Oldest Known Life			2500
Archean	Oldest Known Rocks			3900
Hadear	Age of the Earth			4600

Over the past 150 years detailed studies of rocks throughout the world based on stratigraphic, paleontologic, and correlation studies have allowed geologists to correlate rock units throughout the world and break them into time stratigraphic units. The result is the geologic column, which breaks relative geologic time into units of known relative age. Note that the geologic column was established and fairly well known before geologists had a means of determining absolute ages. Thus in the geologic column shown, the absolute ages in the far right-hand column were not known until recently.

Absolute Geologic Time

Although geologists can easily establish relative ages of rocks based on the principles of stratigraphy, knowing how much time a geologic Eon, Era, Period, or Epoch represents is a more difficult problem without having knowledge of absolute ages of rocks. In the early years of geology, many attempts were made to establish some measure of absolute geologic time. Age of Earth estimated on the basis of how long it would take the oceans to obtain their present salt content. Assumes that we know the rate at which the salts (Na, Cl, Ca, and CO3 ions) are input into the oceans by rivers, and assumes that we know the rate at which these salts are removed by chemical precipitation. Calculations in 1889 gave estimate for the age of the Earth of 90 million years.

Age of Earth estimated from time required to cool from an initially molten state. Assumptions include, the initial temperature of the Earth when it formed, the present temperature throughout the interior of the Earth, and that there are no internal sources of heat. Calculations gave estimate of 100 million years for the age of the Earth. In 1896 radioactivity was discovered, and it was soon learned that radioactive decay occurs at a constant rate throughout time. With this discovery, Radiometric dating techniques became possible, and gave us a means of measuring absolute geologic time.

Radiometric Dating

Radiometric dating relies on the fact that there are different types of isotopes.

Radioactive Isotopes: Isotopes (parent isotopes) that spontaneously decay at a constant rate to another isotope.

Radiogenic Isotopes: Isotopes that are formed by radioactive decay (daughter isotopes).

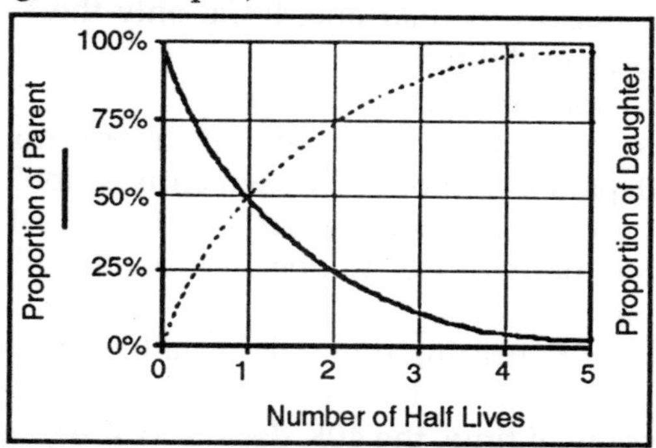

The rate at which radioactive isotopes decay is often stated as the half-life of the isotope (t1/2). The half-life is the amount of time it takes for one half of the initial amount of the parent, radioactive isotope, to decay to the daughter isotope. Thus, if we start out with 1 gram of the parent isotope, after the passage of 1 half-life there will be 0.5 gram of the parent isotope left.After the passage of two half-lives only 0.25 gram will remain, and after 3 half lives only 0.125 will remain etc.

Some examples of isotope systems used to date geologic materials.

Parent	Daughter	t1/2	Useful Range	Type of Material
238U	206Pb	4.5 b.y	>10 million years	Igneous Rocks and Minerals
235U	207Pb	710 m.y		
232Th	208Pb	14 b.y		
40K	40Ar & 40Ca	1.3 b.y	>10,000 years	
87Rb	87Sr	47 b.y	>10 million years	
14C	14N	5,730 y	100 - 70,000 years	Organic Material

Potassium: Argon (K-Ar) Dating In nature there are three isotopes of potassium:

- 39*K*: non-radioactive (stable)
- 40*K*: radioactive with a half life of 1.3 billion years, 40K decays to 40Ar and 40Ca, only the K-Ar branch is used in dating.
- 41*K*: non-radioactive (stable)
- K is an element that goes into many minerals, like feldspars and biotite. Ar, which is a noble gas, does not go into minerals when they first crystallize from a magma because Ar does not bond with any other atom.
- *When a K*: bearing mineral crystallizes from a magma it will contain K, but will not contain Ar. With passage of time, the 40K decays to 40Ar, but the 40Ar is now trapped in the crystal structure where the 40K once was.
- Thus, by measuring the amount of 40K and 40Ar now present in the mineral, we can determine how many half lives have passed since the igneous rock crystallized, and thus know the absolute age of the rock.

Radiocarbon (14C) Dating

Radiocarbon dating is different than the other methods of dating because it cannot be used to directly date rocks, but can only be used to date organic material produced by once living organisms.

- 14C is continually being produced in the Earth's upper atmosphere by bombardment of 14N by cosmic rays. Thus the ratio of 14C to 14N in the Earth's atmosphere is constant.
- Living organisms continually exchange Carbon and Nitrogen with the atmosphere by breathing, feeding, and photosynthesis. Thus, so long as the organism is alive, it will have the same ratio of 14C to 14N as the atmosphere.
- When an organism dies, the 14C decays back to 14N,

with a half-life of 5,730 years. Measuring the amount of 14C in this dead material thus enables the determination of the time elapsed since the organism died.

- Radiocarbon dates are obtained from such things as bones, teeth, charcoal, fossilized wood, and shells.
- Because of the short half-life of 14C, it is only used to date materials younger than about 70,000 years.

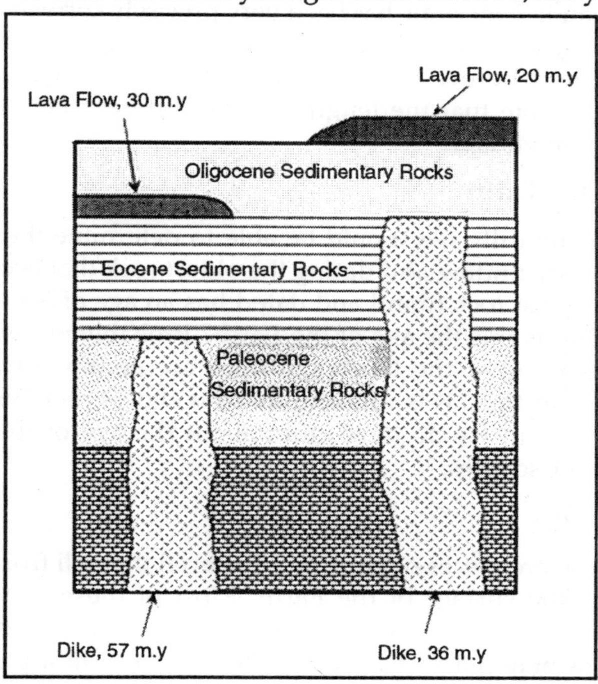

Absolute Dating and Geologic Time Scale

Using the methods of absolute dating, and cross-cutting relationships of igneous rocks, geologists have been able to establish the absolute times for the geologic column. For example, imagine some cross section such as that shown here.From the cross-cutting relationships and stratigraphy we can determine that:

- The Oligocene rocks are younger than the 30 m.y old lava flow and older than the 20 m.y. old lava flow.

- The Eocene rocks are older than the 57 m.y. old dike and younger than the 36 m.y. old dike that cuts through them.
- The Paleocene rocks are older than both the 36 m.y. old dike and the 57 m.y. old dike (thus the Paleocene is younger than 57 m.y.

By examining relationships like these all over the world, the Geologic Time scale has been very precisely correlated with the Geologic Column. but, because the geologic column was established before radiometric dating techniques were available, note that the lengths of the different Periods and Epochs are variable

Age of the Earth

Theoretically we should be able to determine the age of the Earth by finding and dating the oldest rock that occurs. So far, the oldest rock found and dated has an age of 3.96 billion years. But, is this the age of the Earth? Probably not, because rocks exposed at the Earth's surface are continually being eroded, and thus, it is unlikely that the oldest rock will ever be found. But, we do have clues about the age of the Earth from other sources:

Meteorites

These are pieces of planetary material that fall from outer space to the surface of the Earth. Most of these meteorites appear to have come from within our solar system and either represent material that never condensed to form a planet or was once in a planet that has since disintegrated. The ages of the most primitive meteorites all cluster around 4.6 billion years.

Moon Rocks

The only other planetary body in our solar system that we have samples of are moon rocks . The ages obtained on Moon rocks are all within the range between 4.0 and 4.6 billion years. Thus the solar system and the Earth must be at least 4.6 billion years old

Chapter 5

Earthquake

EARTHQUAKE CAUSES

CAUSES OF EARTHQUAKES

When the earth shakes due to the movements of plates below the earth's crust it is known as Earthquake. Earthquakes are natural disasters, which kill thousands of people in an instant and can destroy cities and countries- human habitation across miles. The vibration during an earthquake has the potential to wreak havoc and destruction, which is beyond imagination. The destruction is the maximum near the epicentre, the place from where the vibrations arise and spread.

Of late there have been many such natural disasters, which can be associated with earthquakes like the Tsunami that hit the shores of India, Thailand and razed down buildings and annihilated many lives. It seemed like Mother Nature was avenging herself on us who have used all her endowments to the fullest extent without caring to rejuvenate them. Here we would try to find out the causes of earthquakes so that we can all contribute to the prevention of such things in our own small ways. Individual awareness would definitely lead to mass awareness.

If seen broadly we can say that earthquakes are caused due to two major reasons. The first reason is the eruption of volcanoes, which are sudden, and as is known volcanoes are seat of inner disturbance and can effect the plates which is the second cause of earthquakes. Earthquakes are caused due to disturbance in the movement of plates, which again can be caused due to various reasons like under crust waves or cracks in the plates.

The Herd of Elephants Theory

Have you ever felt the ground shake as a herd of elephants stampeded by? Has your mom ever told you that you sound (and feel!) like a herd of elephants because you are making the whole house shake? Is that an earthquake? Maybe its just a housequake. Have you ever felt your house shake when a truck drives by? Well, that is a very local earthquake. In all these cases, the earth shakes in response to a local shock. These shakes would show up on a seismograph. In fact, they do show up on seismographs and scientists have to know how to tell the difference between a big truck going by outside (or a herd of elephants stampeding down the hall) and a large earthquake halfway around the world.

The Nuclear Explosion Theory

On a larger scale an explosion can cause the earth to shake for a considerable distance. Scientists use seismographs to monitor nuclear tests. People in Las Vegas could feel the shaking caused by underground nuclear tests in the desert miles away. The government analyses the shock waves (earthquakes) produced by nuclear explosions to study the effects of nuclear tests and to monitor tests elsewhere in the world.

The Extraterrestrial, or Meteor, Theory

Every day tiny meteors hit the earth, as we move through space. The vast majority of them burn up in the atmosphere, leaving no more trace than a shooting star across the sky. Once in a while, a meteorite will reach the surface of the earth. Very rarely a great meteorite will hit, causing the ground to shake and creating a large crater. The Meteor Crater in Arizona is an excellent example of this type of crater. Imagine how the ground shook for miles around when it was formed!

The moon is full of meteor craters that we can see because they have not eroded away. The earth also has been struck many times over its history. Erosion by wind and rain wear down the craters so we can't see most of them anymore. Scientists studying the earth have found traces of many meteor impacts around the world. Each impact creates an earthquake.

Volcanic Eruptions

When volcanoes erupt it is because the molten magma under the crust of the earth is under enormous pressure and to release that pressure it looks for an opening and exerts pressure on the earth's crust and the plate in turn. A place, which is the seat of an active volcano, is often prone to earthquakes as well because the pressure that is exerted by the magma exceeds the limit these plates move and that causes earthquakes.

Earthquakes are also caused after a volcanic eruption since the eruption also leads to a disturbance in the position of plates, which either move further or resettle and can result into severe or light tremors.

The excessive exploitation of earth's resources for our own benefits like building dams to store large volumes of water and blasting rocks and mountains to build bridges and roads is also the reason behind such natural disruptions.

VOLCANOES

Earthquakes are one of the indicators of increased volcanic activity leading up to an eruption. As magma forces its way up into a volcano, it pushes aside the rocks in its way, causing bulges in the ground and a flurry of earthquakes. Scientists studying volcanoes watch for an increase in earthquakes to tell them that an eruption may be on the way.

Using this and other measures of volcanic activity, they have been able to warn residents to evacuate before eruptions. Although they still cannot predict eruptions with absolute certainty, they are learning more all the time about what to look for to make better predictions. Much of this knowledge comes from studying volcanic earthquakes.

Plate Tectonic Theory

The outer layer of the earth is divided into many sections known as plates, which are floating on the molten magma beneath the earth's crust. Now the movement of these plates is determined by the convection current in the molten magma. The heat makes these plates rise and vice versa. Therefore after intervals there are plates that get submerged in the molten magma and there plates that rise upwards and at times even new crust is formed from the molten magma which in turn forms a new plate until it connects itself with the already existing ones.

At times these plates and can be pushed up to form mountains and hills and the movement is so slow that it is really hard to comprehend that there is any movement at all. The movement and the results come out to be visible suddenly. Now these plates are the bases on which the continents stand and when these plates move the continents also move. Most of the earthquakes occur on the edges of the plates where a plate is under one or across. This movement disrupts the balance and position of all plates, which leads to tremors, which are called earthquakes.

PLATE TECTONICS

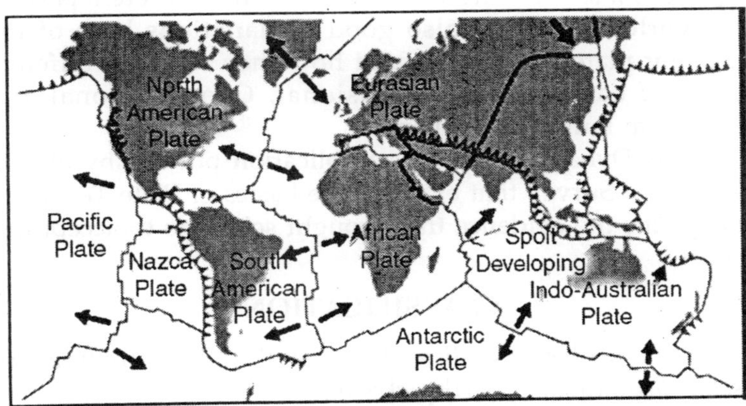

Most Earthquakes are caused by Plate Tectonics. The earth's crust consists of a number of sections or plates that float on the molten rock of the mantle. These plates move on convection currents caused by heat rising from the centre of the earth. The hot magma rises and spreads out on the surface, creating new crust.

The crust spreads out forming a new plate until it meets another plate. One of the plates will be pushed down into the interior of the earth and reabsorbed into the mantle. Plates can also be compressed to push up mountains when they collide or move sideways along transform faults.

The plates are the Earth's crust that float on the molten rock in the centre of the Earth. Most of the inside of the Earth is so hot that the rock melts.

Just as a pot of hot chocolate on the stove will bubble as it is heated; the molten rock, or magma, very slowly bubbles up in great currents under the surface of the Earth. The crust that floats on the magma moves with it, like the skin that might form on the hot chocolate. The Plates are just pieces of the crust.

The part that makes it hard to understand is that it all moves so slowly. Even though the magma is very hot it is also very thick and under tremendous pressure in the middle of the Earth. So it moves only a few centimetres a year. Over millions of years that adds up to a lot of movement. More information from the University of Nevada at Reno explains Plate Tectonics with examples from different parts of the world. There are also good explanations here of the different plate movements and how they result in different kinds of earthquakes. Extensional, Compressional and Transform faults are explained.

This Dynamic Earth is a publication put out by the US Geological Survey that explains the basics of Plate Tectonics. It explains the evidence that brought scientists to accept this theory.

CLASSIFICATION

In the Erdbebenstudien (1878) HOERNES introduced his useful classification of earthquakes in:

- Rock-fall earthquakes
- Volcanic earthquakes and
- Tectonic (afterwards dislocation)

- Earthquakes.

Realising that rock: fall earthquakes are infrequent and volcanic earth-quakes local, he holds that the most numerous, as well as the greatest, of all earthquakes are the products of mountain-formation, and are due to displacements along peripheral and radial fractures of great mountain-chains.

The earthquake occuring in Japan as first reported is a 7.8 quake. This is considerd a Major quake.

The Classification of quakes are as follows:

- Micro 0-3
- Minor 3-3.9
- Moderate 4-4.9
- Strong 6-6.9
- Major 7-7.9

GREAT 8 AND HIGHER

Are you prepared for a disaster? Get prepared by attending local CERT (Community Emergency Responsiveness Training) and checking with your local Fire Dept. as to classes in your region. The best gift you can give your family this holiday season is to learn how to respond in an emergency. You will learn how to survive Earthquakes, Wildfires, Floods, Hurricanes, Tornado's Blizzards and other disasters common in your local area. The training is geared to known disasters in your area.

Classification of Earthquakes

Category	Magnitude on Richter Scale
Slight	Upto 4.9
Moderate	5.0 to 6.9
Great	7.0 to 7.9
Very Great	8.0 and more

Epicentre

It is the point on the surface of the earth, vertically above the place of origin (hypocentre) of an earthquake. This point is expressed by its geographical Coordinates in terms of latitude and longitude.

Hypocentre or Focus

It is the point within the earth, from where seismic waves originate. Focal depth is the vertical distance between the Hypocentre (Focus) and Epicentre.

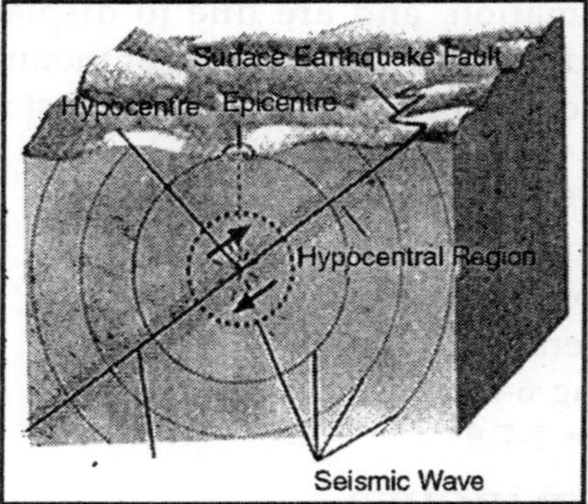

Magnitude

It is a quantity to measure the size of an earthquake and is independent of the place of the observation.

Richter Scale

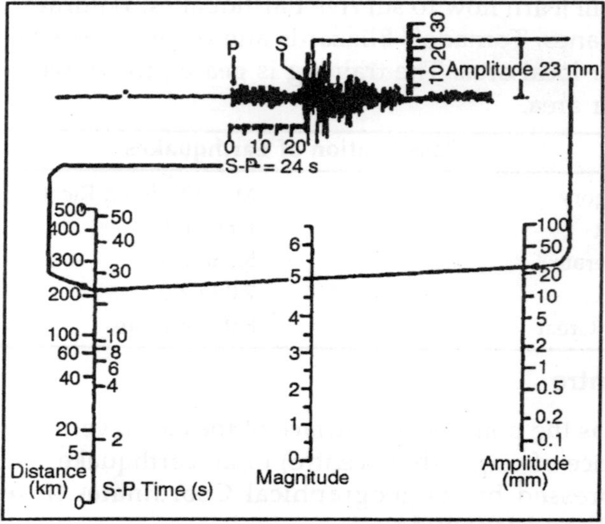

The local magnitude is defined as the logarithm of the maximum amplitude measured in microns on a seismogram written by Wood-Anderson seismograph with free period of 0.8 second, magnification of 2,800, damping factor of 0.8 calculated to be at a distance of 100 kms. The relative size of events is calculated by comparison to a reference event of $M_L=0$,using the formula, $M_L=\log A-\log A_0$ where A is the maximum trace amplitude in micrometre recorded on a standard seismograph and A_0 is a standard value which is a function of epicentral distance (Δ) in kilometres.

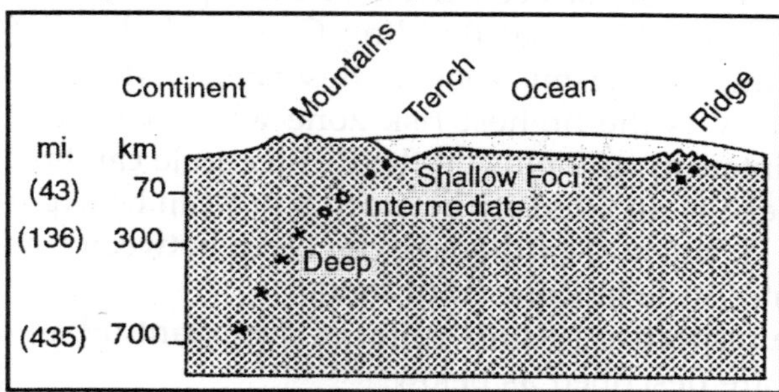

Fig. Depth of Foci.

SEISMIC ZONES OF INDIA

ZONE MAP

According to Seismic Zone Map of India identified in the UEVRP,which have populations exceeding half a million in earthquake Zones III, IV, V.

Zone III

Ahemdabad, Vadodara, Rajkot, Bhavnagar, Surat, Mumbai, Agra, Bhiwandi, nashik, Kanpur Pune, Bhubneshwar, Cuttack, Asansol, Kochi Kolkata, Varanasi, Bareilly, Lucknow, Indore, Jabalpur, Vijaywada, Dhanwad, Chennai, Coimbatore, Manglore, Kozhakode,Trivandrum.

Zone IV

Dehradun, New Delhi, Jamunanagar, Patna, Meerut, Jammu, Amristar,Jalandhar.

Zone V

Guwahati and Srinagar.

Earthquakes occur due to movements along faults that have evolved through geological and tectonic processes. Often they occur without any prior warning and are, therefore, unpredictable. The large area of India is prone to earthquake. The construction of earthquake resistant building is the only solution for safeguarding our urban centres from the menace of earthquakes.

On the basis of occurrence of earthquakes of different intensities, the National Building Code of India divide the country into five seismic zones as described on the above map. The zone V is the highest risk zone where earthquakes of having intensity of 9 plus on the Richter scale can take place. Earthquake of an intensity between 8 to 9 can be experienced in Zone IV whereas earthquake can occur between 6 and 8 on the Richter scale in Zone III of India.

The respective seismic zones of important urban centres of India are described as below.

Town	Seismic Zone	Town	Seismic Zone
Agra	III	Jorhat	V
Ahmadabad	III	Kanpur	III
Ajmer	I	Kathmandu	V
Allahabad	II	Kohima	V
Almora	IV	Kurnool	I
Ambala	IV	Lucknow	III
Amritsar	IV	Ludhiana	IV
Asansol	III	Madras	II
Aurangabad	I	Madurai	II
Bahraich	IV	Mandi	IV
Bangalore	I	Mangalore	III
Barauni	IV	Monghyre	IV
Bareilly	III	Moradabad	IV
Bhatinda	III	Mysore	I
Bhilai	I	Nagpur	IV
Bhopal	II	Nainital	IV
Bhubaneshwar	III	Nasik	III
Bhuj	V	Nellore	II

Bikaner	III	Punjim	III
Bokaro	III	Patiala	III
Bombay	III	Patna	IV
Burdwan	III	Pilibhit	IV
Calcutta	III	Pondicherry	II
Calicut	III	Pune	III
Chandigarh	IV	Rajpur	I
Chitradurga	I	Rajkot	III
Coimbatore	III	Ranchi	II
Cuttack	III	Roorkee	IV
Darbhanga	V	Rourkela	I
Darjeeling	IV	Sadiya	V
Dehra Dun	IV	Shimla	IV
Delhi	IV	Sironj	I
Durgapur	III	Srinagar	V
Gangtok	IV	Surat	III
Guwahati	V	Tezpur	V
Gaya	III	Tanjavur	II
Gorakhpur	IV	Tiruchchirappalli	II
Hyderabad	I	Trivandrum	III
Imphal	V	Udaipur	II
Jabalpur	III	Vadodara	III
Jaipur	II	Varanasi	III
Jamshedpur	II	Vijayawada	III
Jhansi	I	Vishakhapatanam	II
Jodhpur	I		

The natural disasters like earthquake can not be prevented, but measures are required to be taken to reduce the extent of damage, especially in a vast country like India which is the 2nd largest populated country of the world supported by low level infrastructure and inadequate resources. High levels of risk combined with low levels of coping mechanisms result in major disruptions or loss of lives and livelihood.

The developed countries of the world are adopting new technique of construction of seismic proof buildings whereas under developed countries do not give much attention for the construction of seismic proof buildings due to shortage of resources. With the result world's worst disasters always take place in underdeveloped and poor countries. Disasters cause enormous destruction and human sufferings. The losses due to occurrence of earthquakes reduce the pace of economic development and often lead to depletion of available resources.

GEOLOGICAL CONSIDERATION FOR CONSTRUCTION OF BUILDING

Earthquakes occur due to movements along faults that have evolved through geological and tectonic processes. Often they occur without any prior warning and are, therefore, unpredictable. The large area of India is prone to earthquake. The construction of earthquake resistant building is the only solution for safeguarding our urban centres from the menace of earthquakes.

The natural disasters like earthquake can not be prevented, but measures are required to be taken to reduce the extent of damage, especially in a vast country like India which is the 2nd largest populated country of the world supported by low level infrastructure and inadequate resources. High levels of risk combined with low levels of coping mechanisms result in major disruptions or loss of lives and livelihood.

The developed countries of the world are adopting new technique of construction of seismic proof buildings whereas under developed countries do not give much attention for the construction of seismic proof buildings due to shortage of resources. With the result world's worst disasters always take place in underdeveloped and poor countries. Disasters cause enormous destruction and human sufferings. The losses due to occurrence of earthquakes reduce the pace of economic development and often lead to depletion of available resources.

RECOMMENDATIONS

The earthquake disasters can be averted with the

construction of seismic proof buildings. Each building can be designed in such a way that it may withstand during severest quakes depending on the seismic zone it falls in. "The National Disaster Management

Authority" (NDMA) has made it compulsory for all new constructions to be earthquake-resistant, especially in cities located in seismic zones.

The guidelines have also recommended selective seismic strengthening and retrofitting of existing priority structures located in high-risk areas.

It has also been proven that well maintained buildings have faired better than those in poor condition during and after an earthquake. Thus, maintenance and seismic retrofit are two critical components for the protection of historic buildings in areas of seismic activity.

It makes no sense to retrofit a building without improvements. The subcontinent is sitting on the highly seismic Indian plate, with some major faults lines. In fact there is no seismically safe zone in India. Disasters have left the 800-year-old Qutub Minar with a slight tilt but it has survived several quakes in its lifetime.

DISASTER MANAGEMENT PLAN

The Disaster Management Bill, likely to be presented in the winter session of Parliament, will make it necessary for all states to have a disaster management authority and implement the national disaster plan. "Eventually disaster management is a state's concern and the action plan has ultimately depend upon the state's own concerns and ability to set up institutional and financial.

CREATION OF SPECIAL FORCE AT NATIONAL LEVEL

At the national level, other measures are being planned. Eight battalions of 10,000 soldiers are being trained for being posted to eight different locations and money has also been sanctioned for buying aircraft for their use in cases of emergency.

EMERGENCY OPERATION CENTRE

It is being equipped with state-of-th-art communication links and micro-zonation of 38 cities above 10-lakh population is being attempted in different phases. The micro-zonation of Delhi has just been completed.

In words of Science and Technology Minister Kapil Sibal, the micro-zonation process is the government's effort to take effective measures with proper research to minimize risk to existing buildings in the event of an earthquake. Micro-zonation, he says, will help bring area-wise changes in building bylaws to ensure quake resistant measures in the structural designs of high rises to minimize the risk of heavy damage and loss of life in event of an earthquake.

While the government is attempting a paradigm shift in the disaster management from relief and rehabilitation to mitigation and prevention, to make it successful will eventually depend upon the civil society.

PROJECTS IN SEISMIC AREAS

SEISMIC PROJECTS GET NEW SCRUTINY

It's perhaps an understatement to say the last few years of impressive commodity prices and all-out activity in the oil and gas industry have been exciting for everyone involved in the business.

And although volatility is a hallmark of oil and gas prices, the most recent, unusually vicious freefall in prices seemingly was not on anyone's radar screen. "Those of us with gray hair have seen it all," said Steve Mitchell, vice president and division manager at Fairfield Industries. "We saw it go from $70 to $10 pretty quickly, but not as quickly as this particular fall came. "Most everyone wasn't looking over their shoulder enough, because we got slammed in the back with this one," Mitchell added. "Plus, there's the news media blitz of recession, recession, recession, making you think you'd better hunker down before you want to because the market's down," he said, "so deciding where you need to conserve can become a very delicate line."

Still Rolling

Apache Egypt, working with geophysical services provider CGG Ardiseis, utilized 12 vibrator trucks to acquire an unparalled 678 vibe points per hour on this project. Despite current conditions of uncertainty today, geophysical companies are keeping busy on projects that will be needed tomorrow.

RIDING THE CYCLE – AGAIN

But this is a business rife with risk takers who have become adept at weathering the down times – and equally adept at taking advantage of these troughs to prepare for better days ahead. "You naturally go into a survival mode, knowing that budgets and projects are going to be cut," Mitchell said, "and that the service contractors have gotta dance.

"Yet by the same token, as you ax, cut, conserve cash you know it will come back," he said. "And the one thing you have to do is keep the technology going – not just RandD but getting the product to market – so you're not a has-been when it does come back. "Technology is such a buzzword, but what's key is technology that can be readily seen to save money for both the contractor and the client," Mitchell said. "When the end-user oil company can see it will cost less yet give the same quality data or better, that's when you've got a winner."

Mitchell noted the healthy companies that exercised discipline during the recent high-flying times are continuing to develop new technologies despite the current slowdown in the industry and the global recession. In addition to the contractors, certain EandP companies are doing their part in designing and developing new tech applications with an eye to the future.

"We know the business is cyclical and will turn around," said AAPG member Mike Bahorich, executive vice president for exploration and production technology at Apache Corp. "And we know there are a lot of things the industry needs to do to be more successful when it does turn around – and I'm absolutely comfortable the industry will come out of this. "What happens when times get tough, people are forced to innovate," Bahorich noted. "As the saying goes, necessity is the mother of invention, so as we have lower prices we're forced to come up with new methods for doing things more efficiently to produce oil and gas at a price the market will bear."

THE SILVER LINING

Industry downturns can actually be advantageous for certain companies. An example is TGS, which is 95 per cent-focused on multi-client seismic, using a business model that entails putting together multi-client projects and then chartering other companies to acquire the data. "In down times like now, the oil companies will start putting downward pressure on day rates for geophysical vessels," said AAPG member Robert Hobbs, COO at TGS. "Because we're only taking charters for these vessels for individual projects or a short-term string of projects, we can take advantage of these lower rates.

"Because of this very flexible business model, we look at times like this as opportunity to gain market share," Hobbs noted. "We can put together multi-client projects at very efficient rates by taking advantage of the lower vessel costs." "We're careful about evaluating new projects," Hobbs emphasized. "Part of that is thinking ahead to when there's an upturn. "We want to make sure projects we acquire are in areas we believe will see enhanced EandP activity and, therefore, enhanced sales when things pick up," he added.

There are a number of innovations currently in progress

that may increase efficiency in the industry, according to Bahorich. for example, advances in horizontal drilling.

"There are new technologies out there that may enable a much larger number of frac stages in a single horizontal lateral," Bahorich said. "Once you've drilled a horizontal well and put in a number of frac stages in that well, the net developing cost for one additional frac stage is very low because you have the horizontal drilled and all the infrastructure is in. "So it's just the amount of money needed to put one set of perfs and one additional pump job of proppant and fluid,"

EYEING THE INNOVATIONS

Regarding the geosciences side of the industry, Bahorich views high density seismic on land to be one of the most exciting technologies to emerge. He noted the combination of very high-density sources and very high-density receivers is "just around the corner." The result will be still-more improved, more cost-effective imaging.

Marine seismic is one focus of Apache's technology expertise. "We have a patent pending on a new method for shooting marine seismic," Bahorich said. "It's a new method for designing infill that involves a unique shooting geometry.

"It saves 10 to 20 per cent of the cost and gives you equivalent data," he said. "Considering that seismic boats can run a half million dollars a day, savings of this magnitude can be quite significant."

The company also has a patent pending on a logging technology for exploration of unconventional reservoirs. The technology solves the problem for certain zones where there's no log response associated with pay. It enables the operator to see gas pay where, in some instances, it's invisible with traditional logs.

Bahorich emphasized Apache develops ideas in-house and works with outside partners to develop them. Besides spurring technical innovations, industry downturns offer numerous opportunities for large companies as well as the not-so-large to gain muscle by acquiring assets and/or whole companies at attractive prices. Apache appears well positioned to take advantage of the situation given that it has close to a cool $2 billion sitting in its coffers, according to president and CEO Steve Farris. "We're very interested in making significant asset acquisitions, both domestic and international," Bahorich said, noting the company typically buys assets as opposed to companies.

LAY OFF THE LAYOFFS

One of the dark holes of previous downturns in the industry has been that old bugaboo: layoffs.

There's optimism this time may be different given the major effort to staff-up over the last few years after letting so many employees go during past rough patches in the industry – the late '90s being the most recent trough. To date, there's only been a trickle of announced cuts in personnel. Most companies appear to be maintaining the status quo or – in some cases – beefing up staff to be prepared for a rebound in commodity prices. Devon, for instance, is said to be expanding its U.S. intern programme this year by 15 per cent.

Brazilian NOC giant Petrobras, which is sitting on billions of barrels of newly- discovered oil offshore, announced it's hiring personnel and cutting costs in other areas.

Apache has decided to avoid layoffs at this point, according to Bahorich. This mirrors the path the company took in 1998 when its peers were downsizing staff. In fact, he related they went to such extremes to cut costs via other means they actually followed one employee's advice to get rid of all styrofoam cups in the building. "We didn't think it would save much," Bahorich said, "but we were looking at every possible cost savings. "Right after that, the turnaround came," he said, "and we were back in business."

Despite the general consensus that the current slump is temporary, Mitchell noted he thinks the business is facing a tough year – or perhaps slightly more. "After that I think all of a sudden, it's gonna be like it was in the '90s," he said. "We'd all been beaten up so long, and all of a sudden we turned around and it's getting better."

Chapter 6

Landslides

CONCEPT

Landslides are simply defined as the mass movement of rock, debris or earth down a slope and have come to include a broad range of motions whereby falling, sliding and flowing under the influence of gravity dislodges earth material. They often take place in conjunction with earthquakes, floods and volcanoes. At times, prolonged rainfall causing heavy block the flow or river for quite some time. The formation of river blocks can cause havoc to the settlements downstream on it's bursting.

In the hilly terrain of India including the Himalayas, landslides have been a major and widely spread natural disaster the often strike life and property and occupy a position of major concern

One of the worst tragedies took place at Malpa Uttarkhand (UP) on 11th and 17th August 1998 when nearly 380 people were killed when massive landslides washed away the entire village. This included 60 pilgrims going to Lake Mansarovar in Tibet. Consequently various land reform measures have been initiated as mitigation measures.

The two regions most vulnerable to landslides are the Himalayas and the Western Ghats. The Himalayas mountain belt comprise of tectonically unstable younger geological formations subjected to severe seismic activity. The Western Ghats and nilgiris are geologically stable but have uplifted plateau margins influenced by neo- tectonic activity. Compared to Western Ghats region, the slides in the

Himalayas region are huge and massive and in most cases the overburden along with the underlying litho logy is displaced during sliding particularly due to the seismic factor.

Landslides are simply defined as the mass movement of rock, debris or earth down a slope and have come to include a broad range of motions whereby falling, sliding and flowing under the influence of gravity dislodges earth material. They often take place in conjunction with earthquakes, floods and volcanoes. At times, prolonged rainfall causing heavy block the flow or river for quite some time. The formation of river blocks can cause havoc to the settlements downstream on it's bursting.

In the hilly terrain of India including the Himalayas, landslides have been a major and widely spread natural disaster the often strike life and property and occupy a position of major concern

One of the worst tragedies took place at Malpa Uttarkhand (UP) on 11th and 17th August 1998 when nearly 380 people were killed when massive landslides washed away the entire village. This included 60 pilgrims going to Lake Mansarovar in Tibet. Consequently various land reform measures have been initiated as mitigation measures.

The two regions most vulnerable to landslides are the Himalayas and the Western Ghats. The Himalayas mountain belt comprise of tectonically unstable younger geological formations subjected to severe seismic activity. The Western Ghats and nilgiris are geologically stable but have uplifted plateau margins influenced by neo- tectonic activity. Compared to Western Ghats region, the slides in the Himalayas region are huge and massive and in most cases the overburden along with the underlying litho logy is displaced during sliding particularly due to the seismic factor.

INCIDENCES OF LANDSLIDES IN INDIA

Region	Incidences of Landslides
Himalayas	High to very high
North-eastern Hills	High
Western Ghats and the Nilgiris	Modern to high

Cattle lost	Low
Vindhayachal	Low

Landslides Zonation Mopping is a modern method to identify landslides prone areas and has been in use in India since 1980s

The major parametres that call for evaluation are as follows:

- *Slope*:Magnitude, length and Direction
- Soil thickness
- Relative relief
- Land use
- *Drainage*: pattern and density
- Landslide affected population

CAUSES

CAUSES OF LANDSLIDES

Landslides can be caused by poor ground conditions, geomorphic phenomena, and natural physical forces and quite often due to heavy spells of rainfall coupled with impeded drainage.

CHECKLIST OF CAUSES

- Weak, sensitivity, or weathered materials
- Adverse ground structure (joints, fissures etc.)
- Physical property variation (permeability, plasticity etc)

MORPHOLOGICAL CAUSES

- Ground uplift (volcanic, tectonic etc)
- Erosion (wind, water)
- Scour4. Deposition loading in the slope crest5. Vegetation removal (by forest fire, drought etc)

PHYSICAL CAUSES

- Prolonged precipitation
- Rapid draw- down
- Earthquake
- Volcanic eruption

- Thawing
- Shrink and swell
- Artesian pressure

SETTLEMENT POLICY

Drawing upon the kerala study in parts of Western Ghats (referred to earlier), it has been felt that while permanent settlement should be avoided in high-risk zones, site selection even in moderately safe zones, especially in plateau edge regions should be made with caution. Diversion of stream channels in upper slopes, especially above settlement should strictly disallow.

Adequate provision should made to ensure drainage of storm water away from the high sloping terrain so as to reduce over saturation, Any contour bounding, or terracing adopted for seasonal cultivation or initiation of plantations in slopes of >16o above settlement should have sufficient provision for storm water drainage. Further, in such areas the existing natural drainage channels and hallows are to be meticulously maintained without any attempt at blocking, division or modification.

CLASSIFICATION AND PREVENTIVE MEASURES

MITIGATORY MEASURES

In general the chief mitigatory measures to be adopted for such areas are :

- Drainage correction,
- Proper land use measures,
- Reforestation for the areas occupied by degraded vegetation and
- Creation of awareness among local population.

The most important triggering mechanism for mass movements is the water infiltrating into the overburden during heavy rains and consequent increase in pore pressure within the overburden. When this happens in steep slopes the safety factor of the slope material gets considerably reduced causing it to move down. Hence the natural way of preventing this situation is by reducing infiltration and allowing excess water to move down without hindrance. As such, the first and foremost mitigation measure is drainage correction. This involves maintenance of natural drainage channels both micro and macro in vulnerable slopes.

The universal use of contour bounding for all types of terrain without consideration of the slope, overburden thickness and texture or drainage set- up needs to be controlled especially in the plateau edge regions. It is time to think about alternative and innovations, which are suitable for the terrain, to be set up. It need not be over-emphasized the governmental agencies have a lot to contribute in this field.

Leaving aside the 'critical zones' with settlements could be avoided altogether and which could be preferably used for permanent vegetation, the 'highly unstable zones' generally lie in the upper regions, which are occupied by highly degraded vegetation.

These areas warrant immediate afforestation measures with suitable plant species. The afforestation programme should be properly planned so the little slope modification is done in the process. Bounding of any sort using boulders etc. has to be avoided. The selection of suitable plant species should be such that can with stand the existing stress conditions in this terrain.

GENERAL PREVENTIVE MEASURES

- Avoid a high protein diet. For the average American this means cutting your beef, pork, lamb, fish and cheese consumption at least in half. Excess protein in the diet acidifies the urine and cause excess urinary calcium.
- Drink enough water to cause a urine output of +2000cc (two quarts). Usually about 3 liters or quarts per day.
- Avoid Colas. This is associated with a low urine pH which may cause the urine to form stones.
- Avoid a high salt diet. Excess salt causes excess urinary calcium.

SPECIFIC TREATMENT PROGRAMMES

- Idiopathic (no known associated disease process).
- Uric acid ("gout") related (may produce calcium oxatate or uric acid stones)

  A low purine diet is indicated. Allopurinol will lower plasma uric acid and will usually prevent these stones. Sodium bicarbonate (usually about 1 to 1.5 mEq/kg/day) will raise the urine pH, making uric acid stone formation less likely. The goal is for the urine pH to be between 6 and 6.5.
- Hyperoxaluria Calcium carbonate (Tums) will bind oxalate in the intestines thereby reducing the oxalate excretion in the urine. It may also reduce the acidity of the urine. Both these effects of calcium carbonate in combination with the "General preventive measures" are effective measures in stone prevention.

  One Tums (500mg) four time per day is the recommended dose. It should be taken with food or later in the day when food is in the intestinal track (i.e. when there is food containing oxalate to bind). If taken on an empty stomach it may have a reverse effect by raising the calcium level in the urine.

  In crohn's disease or small intestinal resection Cholestyramine will bind the excess bile reducing the colonic absorption of oxalate.

  Pyridoxine (Vitamin B 6) will reduce the urinary excretion of oxalate.

  A dietary restriction of oxalate containing food is also helpful. High oxalate foods are, rhubarb, peanuts, chocolate, spinach, beets, strawberries, tomatoes and strongly brewed teas.

  Some recommended foods that are low in oxalates and protein but rich in vitamins include brewer's yeast, soybeans, bananas, baked potatoes with skins, watermelon, avocados, and dried cereals (not wheat brain however).
- Urinary track infection related (Struvite or Calcium phosphate Stone)

Identifing the bacteria is important and base the choice of antibiotic on the sensitivities. Prolong treatment is required especial if there are stones present.

- Hypocitraturia (Renal tubular acicdosis, idiopathic or hypokalemic related)

  The goal in treatment is to reduce the acidity of the urine. Potassium citrate in a dose of 1 mEq/kg is a effective therapy and has been shown to reduce the incidents of new stones.

- Cystine stone
  - Increase urine volume. It is needed, at the usual acidic pH, to dilute the cystine in the urine to 300mg/L. If the cystine excreted in the urine is 900mg per day then 3+liters of urine would need to be produced. This would require about 1 gallon of water/fluid ingestion per day. Unfortunately some with this disorder produce + 1000 mg/day making this dilution difficult to achieve.
  - Increase urine pH to +7.5 making cystine more soluable. Also difficult to ingest enough Sodium bicarbonate and/or potassium citrate..
  - Reduce salt intake
  - Protein restriction
  - Medications

  (a) Penicillamine toxic but low nocturnal dose may be helpful.

  (b) Captopril (50mg TID) only somewhat helpful.

MAGNESIUM DEFICIENCY

Magnesium is a inhibitor of the formation of calcium oxalate crystals in the urine. In a study by Johansson and Backman U in the Journal of the American College of Nutrition a trial was done to determine the effect a magnesium therapy on preventing calium containing kidney stones. 55 patients with recurrent renal calcium stone disease without signs of magnesium deficiency (normal serum magnesium and urinary magnesium) from their "outpatient stone clinic" were treated for up to four years with 500 mg Mg2+, in the form of Mg(OH)2, daily. "The mean stone episode rate before therapy was 0.8 stones/year/patient. Forty-three recurrent renal calcium stone-formers without medical therapy served as controls.

The urinary calcium excretion remained unchanged. The magnesium/calcium ratio in the urine increased and approached a value earlier found in healthy subjects without stone disease. Urinary citrate increased on therapy when analysed after three years of treatment. The mean stone episode rate decreased from 0.8 to 0.08 stones/year on treatment and 85% of the patients remained free of recurrence during follow-up, whereas 59% of the patients in the control group continued their stone formation. Side effects were few." They concluded that "magnesium treatment in renal calcium stone disease is effective with few side effects. No clinical signs of magnesium excess were observed.

Chapter 7

Underground Water

OCCURRENCE

Most groundwater originates from rainfall that has entered the earth. The drawing shows a typical situation with water saturated soils (overburden aquifer) over a bedrock aquifer. In the overburden aquifer, water fills the void space between grains of the soil.

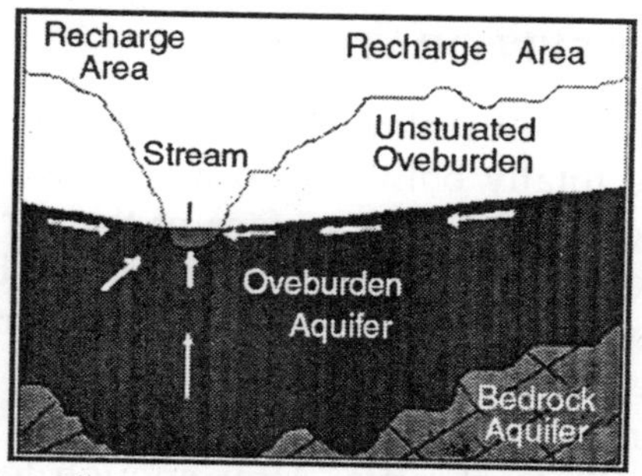

Bedrock aquifers underlie the surface soils (overburden) and overburden aquifers. In the bedrock aquifers, water occurs in fractures and other voids in the bedrock.

Some types of bedrock such as sandstone may also have additional voids (intergranular voids) that are filled with groundwater.

As is the case for surface water, groundwater flows from higher elevations (or pressures) toward lower elevations (or lower pressures).Groundwater flow is usually toward a groundwater discharge area, as shown in the illustration. The stream in the drawing represents a typical groundwater discharge area.

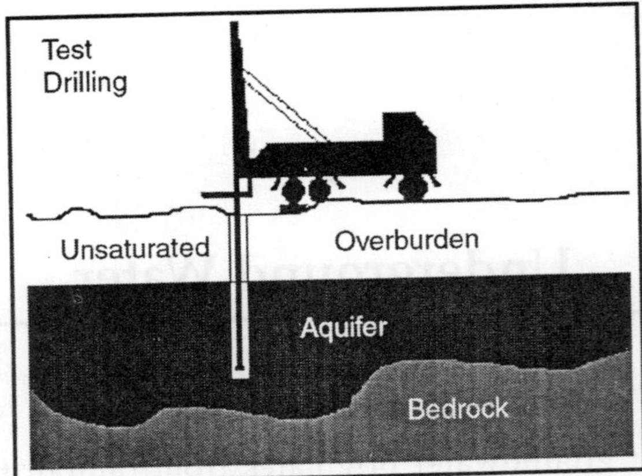

Groundwater pressure, rather than elevation, controls the rate and direction of flow in confined (or artesian) aquifers. Those are aquifers that are isolated under impervious or poorly pervious strata (aquicludes and aquitards).

SUBSURFACE EXPLORATION

The subterranean realm is underground everywhere and most of us are totally oblivious to its existance. In order to know what exists below the surface of the earth, subsurface exploration of one type or another must be done. The illustration on the right shows one type of subsurface exploration, test drilling.

During drilling, samples of the earth may be brought to the surface for us to observe. After enough test drilling has been completed, the structure of the subsurface can be reconstructed. The borings can then be converted into wells by installing pipe with slots or holes, or a screen section to permit water to enter.

Then the depth to water can be measured and groundwater samples can be collected for analysis to investigate flow direction, the composition of the water, and if it contains contaminants.

There are other methods that may be used to explore the subterranean realm. They include excavation and mining, cavern entry and exploration, and indirect methods that may involve geophysical techniques.

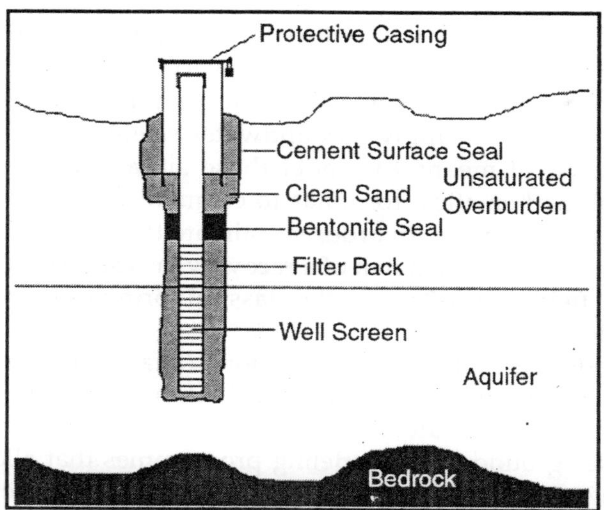

GROUNDWATER MONITORING

Groundwater monitoring is performed in a number of situations with varying objectives. It involves measuring the physical and/or chemical properties of groundwater on a periodic basis. Concentrations of contaminants are frequently monitored to determine if they are increasing, decreasing, or remaining in the same range. Monitoring is also performed at and in the vicinity of water supply sources to determine the quality of the water and trends of indicators of water quality. Groundwater monitoring programmes normally involve an array of monitoring or observation wells or microwells.

The drawing on the right indicates a shallow overburden aquifer monitoring well and some of its basic features. A well of this type makes it possible to measure the groundwater elevation and permits water sampling to test the composition of the groundwater.

GROUNDWATER MODELING

Groundwater Flow Modeling

Groundwater flow modeling is generally used to define the quantity of groundwater available or direction of dissolved contaminant migration. It is also used to define the limits of a capture zone for a contamination recovery well (or well field), or for delineating a water well protection area (or recharge area) for a water supply. Groundwater scientists frequently use analytical means to model these situations using the classical mathematical formulas to estimate the effect on the groundwater surface. In order to estimate the long-term yield and water-level drawdown of a recovery or water supply well (or well field) they often use the classical formulas for making projections.

Modeling by manual calculations (analytical modeling) can be very time-consuming and, therefore expensive. Therefore, groundwater scientists have prepared a number of computer groundwater modeling programmes that allow for a more rapid and efficient assessment of groundwater flow under conditions that may involve the addition of simulated wells and/or simulated sources of recharge in an existing flow field. The models often generate contour maps that illustrate relevant data that are related to groundwater flow. The computer output is usually plotted as groundwater elevation (or artesian pressure) maps for further analysis.

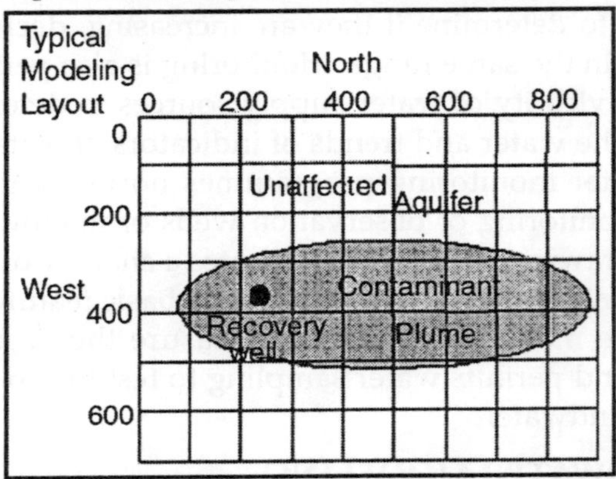

SOLUTE TRANSPORT MODELING

Groundwater Quality

Groundwater quality reflects substances that are dissolved

or suspended in the water. Suspended material is not transported far in most subsurface materials, but it is usually filtered out.

In general, groundwater flow is very slow and depends on the permeability (water transmitting ability) of the subsurface materials, as wells as the hydraulic gradient (slope of the water-table or pressure gradient for artesian conditions).

The rate of groundwater flow is usually measured in feet (or metres) per day or feet (or metres) per year. In some situations where flow is slow it is measured in inches (or centimetres) per year.

Groundwater usually contains higher concentrations of natural dissolved materials than surface water. The materials dissolved in the water usually reflect the composition and solubility of the earth materials (soil or rock) that the groundwater is in contact with and time that it has been in the subsurface.

A number of the activities of man pose threats to water quality. Some of these

Activities include:

- Landfill solid waste disposal
- Liquid waste disposal basins
- Septic waste infiltration systems
- Highway deicing with chemicals (eg. salt)
- Gasoline service stations
- Petroleum bulk storage facilities
- Underground storage tanks
- Many industrial activities
- Livestock feed lots
- Urban stormwater infiltration

The illustration presents an example of how a source of groundwater contamination can pollute millions of gallons of groundwater in an underlying aquifer. The "industrial area" is presented as a typical source of contamination. In the illustration, the groundwater contaminants are volatile organic chemicals (eg. trichloroethene or TCE) that are used as solvents or degreasers in various industrial processes. The degree of contamination is indicated by the concentration of total volatile organics by contours.

Degree of contamination is also illustrated by varying colours in the aquifer. Those closest to the industrial area have the highest concentrations, while those at a distance or upgradient have lower concentrations. The downgradient water supply well is being impacted by the contamination from the industrial area and may have to be shut down until the aquifer is remediated (cleaned up).

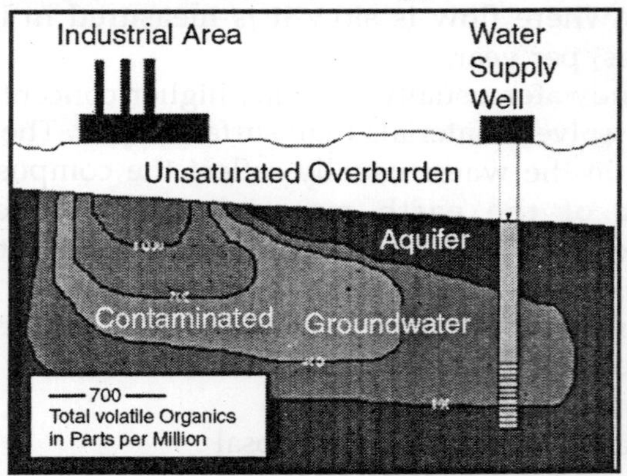

Groundwater scientists have used a number of different computer models to evaluate the transport of various dissolved organic and inorganic compounds in groundwater in a number of geologic situations.

One frequently encountered modeling situation involves the assessment of the distribution of hydrocarbon concentrations around a contaminant source to evaluate various remedial scenarios including removal of the source of contamination, concentration reductions resulting from consumption by microbes, effect of groundwater recovery wells, and recharge sources.

The illustration on the right shows a simple grid layout for solute transport modeling. Such models can be used to predict the time required for aquifer cleanup or for natural concentration reductions by existing processes in the subsurface.

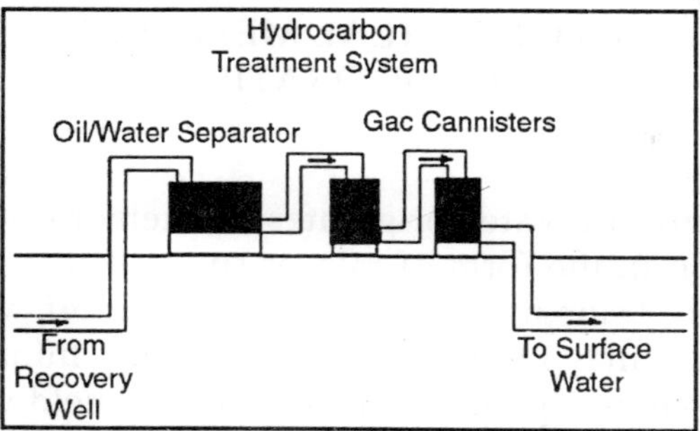

GROUNDWATER REMEDIATION

The illustration below shows a schematic drawing for a treatment system for removing petroleum (gasoline or oil) components from groundwater. If groundwater contamination is identified on a site, and if contaminant concentrations are found above regulatory limits, remedial activities or feasibility studies must be done just to keep a site in compliance. Such activities vary with the contaminant, medium that is contaminated, and surrounding environmental factors.

Common remedial methods include the following:

- Excavation and offsite removal.
- Excavation and onsite treatment
- Groundwater "pump and treat"
- Soil vapour extraction Sparging
- Passive recovery of non-aqueousphase liquid (NAPL)
- Enhanced bioremediation
- Onsite encapsulation
- In-situ onsite treatment

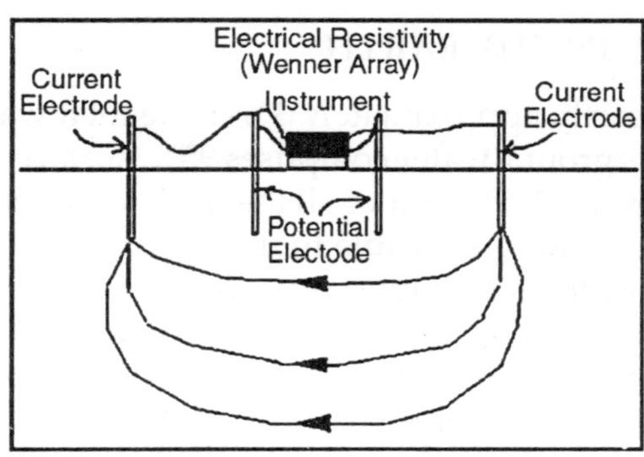

GEOPHYSICAL METHODS OF EXPLORATION

ORIGIN

Most groundwater originates as meteoric water from precipitation in the form of rain or snow. If it is not lost by evaporation, transpiration or to stream runoff, water from these sources may infiltrate into the ground. Initial amounts of water from precipitation onto dry soil are held very tightly as a film on the surfaces and in the micro pores of soil particles in a belt of soil mixture.

At intermediate levels, films of water cover the solid particles, but air is still present in the voids of the soil. This region is called unsaturated zone or zone of aeration, and the water present is vadose water. At lower depths and in presence of adequate amounts of water, all voids are filled to produce a zone of saturation, the upper level of which is the water table. Water present in a zone of saturation is called groundwater.

The porosity and structure of the ground determine the type of aquifer and underground circulation. groundwater may circulate and be stored in the entire geological stratum: this is the case in porous soils such as sand, sandstone and alluvium. It may circulate and be stored in fissures or faults in compact rocks that are not themselves permeable, like most of volcanic and metamorphic rocks. Water trickles through the rocks and circulates because of localized and dispersed fissures. Compact rocks of large fissures or caverns are typical of limestone.

QUANTITY IN THE EARTH

On the earth, approximately 3% of the total water is fresh water. Of this groundwater comprises 95%, surface water 3.5% and soil moisture 1.5%. Out of all the fresh water on the earth, only 0.36% is readily available to use (*Leopold, 1974*).

Groundwater is an important source of water supply. 53% of the population of US receives its water supply from groundwater sources. Groundwater is also a major source of industrial and agricultural uses.

We are withdrawing water from underground aquifers at a faster rate that it can be replenished. Although immense, world's aquifers are not bottomless and in many areas water levels are sinking fast. The water in some aquifers is millennia old and lies beneath what are now some of the driest regions on Earth. Although people have drown water from from springs and wells since the earliest civilizations, in the past 50 years multiplying populations have needed more food and water and the rate of withdrawal has increased drammatically.

In some coastal areas so much fresh water has been withdrawn from aquifers that saltwater has started to intrude, turning well water brackish and unusable.

In some areas the emptying of aquifers has caused serious subsidence or high decreasing of the water table level. Some striking cases are listed here.

USA

- San Joachim Valley, CA - 10 metres
- Phoenix, AZ - more than 1 metre
- Houston - Galveston, TX - 1 metre
- Milwaukee - water table dropped 114 metres by 1976
- Chicago, IL - water table dropped 274 metres by 1979. Partly recovered since then because reduced extraction

Ogallala Aquifer - some wells have run out in Oklahoma, Kansas and in Texas, where the water table has dropped by 30 metres.

Mexico

Mexico City - the city centre has subsided by 7.5 metres since 1950 Ciudad Juarez/El Paso (US border) - the aquifer that supports 1.5 million people is expected to be depleted within 30 years.

Libya

1 billion cubic metres of water per year is being mined by Libya from beneath the Sahara and piped to its farms and cities in the north.

Lebanon

Mining the aquifer beneath Tripoli is leading to an annual deficit of 3.8 million cubic metres.

Yemen

Water table dropping by about 2 metres per year. Wells have been dug 2 km deep without success.

Baluchistan, Pakistan

Water table is dropping by 3.5 metres per year.

Punjab, India and Pakistan

Water table is dropping by 1 metre per year.

North China Plain

water table dropping by 3 metres per year

Groundwater is also affected by water engineering: for decades and centuries, through improper disposal of wastes to the environment and subsurface areas many groundwater have become contaminated. Efforts to protect the quality and quantity of groundwater have been made by cooperation between all government agencies, industrial parties and researchers.

Here you find a map of groundwater distribution on the earth .

AQUIFERS

The blazing sun to dig this hole at the beach. It is a great way to illustrate the concept of how, below a certain depth, the ground, if it is permeable enough to hold water, is saturated with water. The upper surface of this zone of saturation is called the water table.

The saturated zone beneath the water table is called an aquifer, and aquifers are huge storehouses of water. What you are looking at in this picture is a "well" that exposes the water table, with an aquifer beneath it. Ground water is one of our most valuable resource—even though you probably never see it or even realise it is there.

As you may have read, most of the void spaces in the rocks below the water table are filled with water. But rocks have different porosity and permeability characteristics, which means that water does not move around the same way in all rocks below ground.

When a water-bearing rock readily transmits water to wells and springs, it is called an aquifer. Wells can be drilled into the aquifers and water can be pumped out. Precipitation eventually adds water (recharge) into the porous rock of the aquifer.

The rate of recharge is not the same for all aquifers, though, and that must be considered when pumping water from a well. Pumping too much water too fast draws down the water in the aquifer and eventually causes a well to yield less and less water and even run dry. In fact, pumping your well too fast can even cause your neighbour's well to run dry if you both are pumping from the same aquifer.

In the diagram below, you can see how the ground below the water table (the blue area) is saturated with water. The "unsaturated zone" above the water table (the greenish area) still contains water (after all, plants' roots live in this area), but it is not totally saturated with water. You can see this in the two drawings at the bottom of the diagram, which show a close-up of how water is stored in between underground rock particles.

Sometimes the porous rock layers become tilted in the earth. There might be a confining layer of less porous rock both above and below the porous layer. This is an example of a confined aquifer. In this case, the rocks surrounding the aquifer confines the pressure in the porous rock and its water. If a well is drilled into this "pressurized" aquifer, the internal pressure might (depending on the ability of the rock to transport water) be enough to push the water up the well and up to the surface without the aid of a pump, sometimes completely out of the well. This type of well is called artesian. The pressure of water from an artesian well can be quite dramatic.

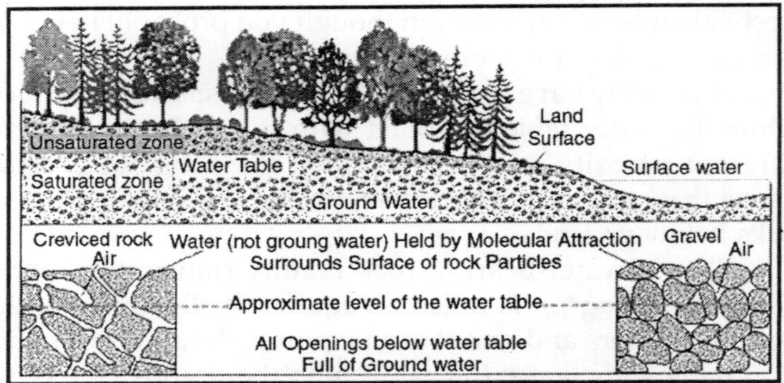

A relationship does not necessarily exist between the water-bearing capacity of rocks and the depth at which they are found. A very dense granite that will yield little or no water to a well may be exposed at the land surface.

Conversely, a porous sandstone, such as the Dakota Sandstone mentioned previously, may lie hundreds or thousands of feet below the land surface and may yield hundreds of gallons per minute of water. Rocks that yield freshwater have been found at depths of more than 6,000 feet, and salty water has come from oil wells at depths of more than 30,000 feet.

On the average, however, the porosity and permeability of rocks decrease as their depth below land surface increases; the pores and cracks in rocks at great depths are closed or greatly reduced in size because of the weight of overlying rocks.

EFFECT OF PUMPING

Under Ground water occurs in the saturated soil and rock below the water table. If the aquifer is shallow enough and permeable enough to allow water to move through it at a rapid-enough rate, then people can drill wells into it and withdraw water.

The level of the water table can naturally change over time due to changes in weather cycles and preciptiation patterns, streamflow and geologic changes, and even human-induced changes, such as the increase in impervious surfaces on the landscape.

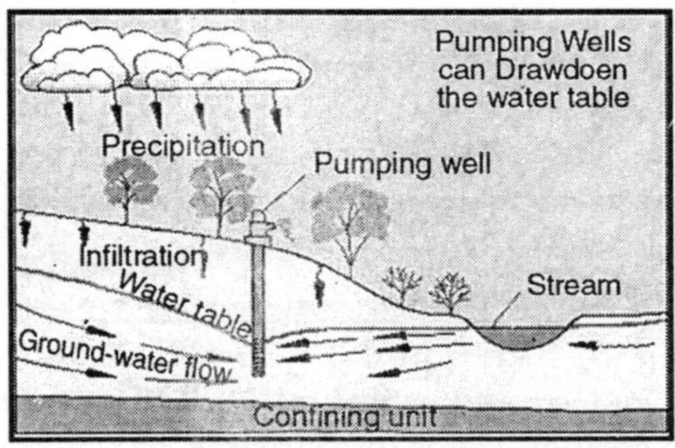

The pumping of wells can have a great deal of influence on water levels below ground, especially in the vicinity of the well, as this diagram shows. If water is withdrawn from the ground at a faster rate that it is replenished, either by infiltration from the surface of from streams, then the water table can become lower, resulting in a "cone of depression" around the well.

Depending on geologic and hydrologic conditions of the aquifer, the impact on the level of the water table can be short-lived or last for decades, and it can fall a small amount or many hundreds of feet. Excessive pumping can lower the water table so much that the wells no longer supply water—they can "go dry."

WATER MOVEMENT IN AQUIFERS

Water movement in aquifers is highly dependent of the pearmeablility of the aquifer material. Permeable material contains interconnected cracks or spaces that are both numerous enough and large enough to allow water to move freely. In some permeable materials groundwater may move several metres in a day; in other places, it moves only a few centimetres in a century. Groundwater moves very slowly through relatively impermeable materials such as clay and shale.

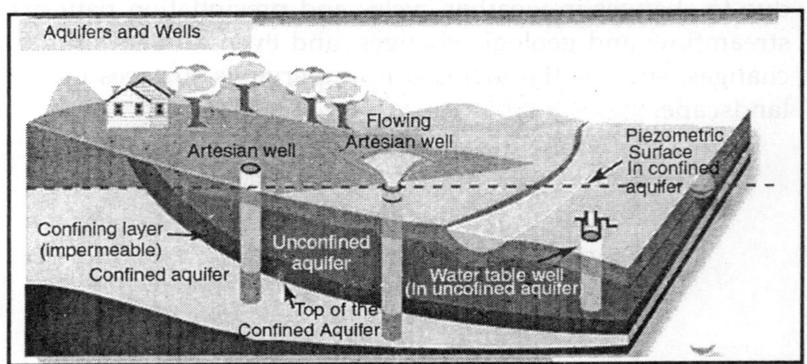

After entering an aquifer, water moves slowly toward lower lying places and eventually is discharged from the aquifer from springs, seeps into streams, or is withdrawn from the ground by wells.

Ground water in aquifers between layers of poorly permeable rock, such as clay or shale, may be confined under pressure. If such a confined aquifer is tapped by a well, water will rise above the top of the aquifer and may even flow from the well onto the land surface. Water confined in this way is said to be under artesian pressure, and the aquifer is called an artesian aquifer.

VISUALIZING ARTESIAN PRESSURE

Here's a little experiment to show you how artesian pressure works. Fill a plastic sandwich baggie with water, put a straw in through the opening, tape the opening around the straw closed, DON'T point the straw towards your teacher, and then squeeze the baggie. Artesian water is pushed out through the straw.

MINING THE AQUIFER

When a natural resource – like iron or gold – is mined, it's usually taken out of the ground and never replaced.

Groundwater is a different type of natural resource. It can be replaced. Gradually, rain that gets trapped in the soil (instead of running off into rivers) will percolate down until it reaches the gravel and sand layers that will hold it underground. The problem is that the process takes a very long time.

Even as the first irrigation wells were being drilled, there were those who realized that the wells would remove water faster than it could be replaced. Irrigators were mining the water in the underground aquifer. But not everyone shared that belief.

Early groundwater irrigators were so impressed by the size of the Ogallala Aquifer that they thought it contained an "inexhaustible" supply.

Some thought that the snows on the Rockies melted and flowed directly into the aquifer, or that a distant arctic glacier was the source. They argued the Ogallala was inexhaustible, so it didn't matter how much water was pumped out, even if some of it was wasted.

After World War II, three factors combined to produce a drastic drop in the water table, particularly in the southern plains state of Texas.

- First, the number of new wells exploded because farmers had money in their pockets and well and pump technology was getting better and cheaper.
- Second, many farmers began to irrigate more than one crop, running the pumps longer each year.
- Third, farmers remembered the drought of the 1930s and feared another.

In some sections of Texas, the drop in the water table reached an astounding 100 feet because of irrigation. By 1949, the Texas Board of Water Engineers issued a dire warning:

"If the present trends of pumping and water-level decline continue, those areas [under pump irrigation] and other parts of the irrigated region will be seriously affected within five to 10 years." Irrigators, on the other hand, were just beginning to feel like their pumps made it possible to farm with a lot more security. They vigorously fought against any suggestion that their use of underground water would have to be regulated. Yet the water table kept dropping.

York County farmer Jim Chenault knows first hand what happens when you mine the aquifer under your land. He has pumped his well dry and is looking for another vein of water. "I'm at the bottom of the hole," Jim laughs, "so I'm hoping that it will start raining. I hate to say it, and the farmer doesn't want to admit the possibility, but there probably needs to be some regulations and some controls on how much water can be pumped." Jim smiles then and says that if his neighbours."

AQUICLUDES

ABANDONED WELL

A well whose use has been permanently discontinued, or a well which is in such disrepair that its continued use for the purpose of obtaining groundwater is impractical or may be a health hazard. Such wells can be a direct "pipeline" for introducing contamination into the groundwater.

AQUICLUDE:

A geologic formation which may contain water but does not transmit it.

AQUIFER/AQUIFER SYSTEM

A water-bearing layer(formation) of rock or sediment capable of yielding sufficient, economical quantities of water to wells. Typically is unconsolidated deposits or sandstone or limestone. DIAGRAM

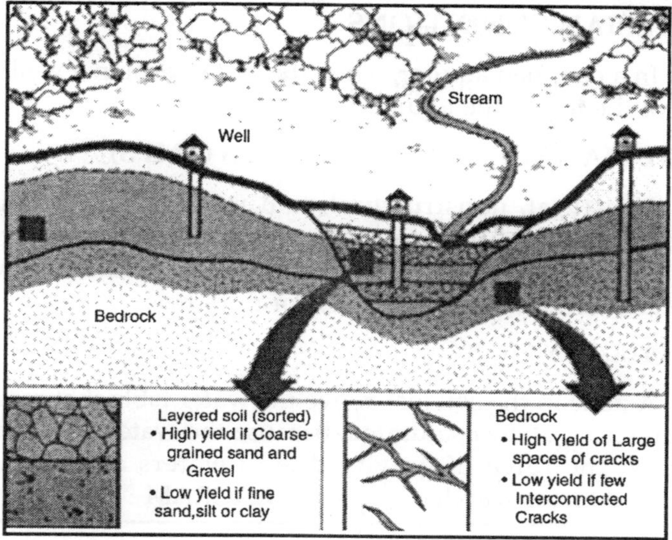

AQUITARD

A saturated but poorly permeable formation that does not yield water freely to a well or a spring. However, an aquitard may transmit appreciable water to or from adjacent aquifers.

AREA OF INFLUENCE

Area surrounding a pumping or recharging well within which the water table or potentiometric surface has been changed due to the well's pumping or recharge. DIAGRAM

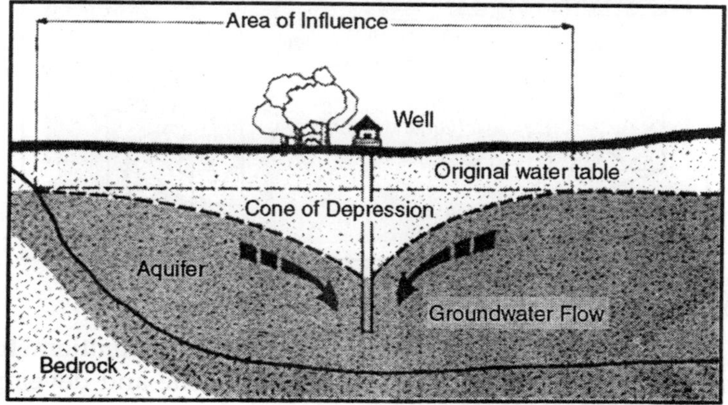

ARTESIAN CONDITIONS

In a confined aquifer, when the water level in a well rises above the top of the aquifer.

BEST MANAGEMENT PRACTICES (BMPS)

Structural, nonstructural, and managerial techniques recognized to be the most effective and practical means to reduce surface water and groundwater contamination while still allowing the productive use of resources.

CAPILLARY FRING

A zone in the soil just above the water table that remains saturated or almost saturated, because the water is drawn up against the force of gravity due to waters adhesive and cohesive forces (capillary action). DIAGRAM

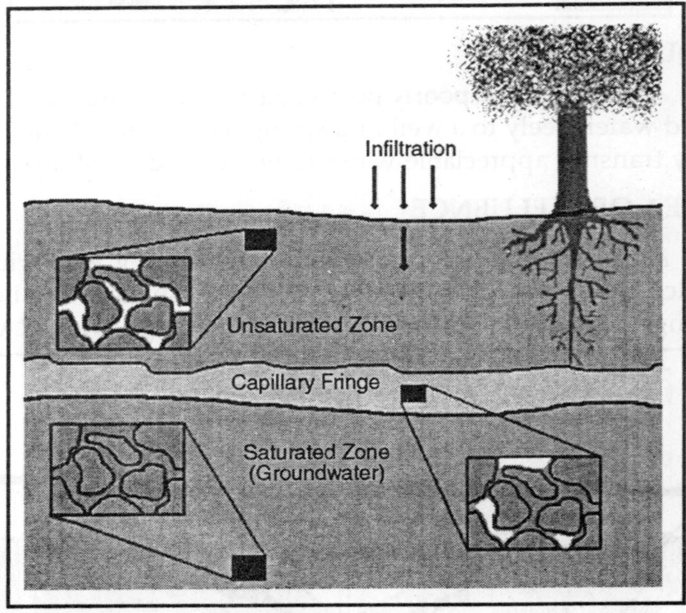

CONE OF DEPRESSION

A depression in the groundwater table or potentiometric surface that has the shape of an inverted cone and develops around a well from which water is being withdrawn. The slopes of the cone become increasingly steep the closer they are to the well. Its trace (perimetre) on the land surface defines the zone of influence of a well. Also called cone of drawdown. DIAGRAM

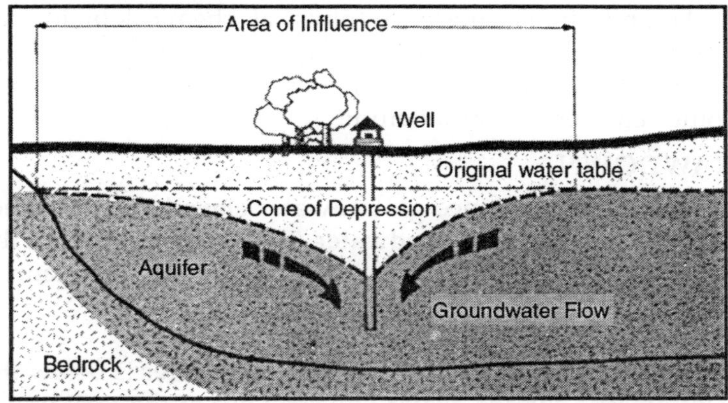

CONFINED AQUIFER

Aquifers that are wedged between layers of relatively impermeable materials and are consequently under pressure. Also know as an artesian aquifer. DIAGRAM

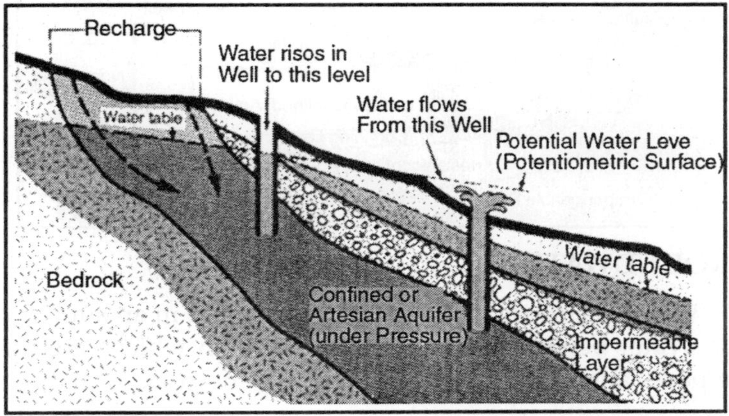

CONFINING LAYER

Geological material through which significant quantities

of water can not move; located below unconfined aquifers, above and below confined aquifers. Also known as a confining bed.

CONTAMINANT

An undesirable substance (physical, chemical, biological, or radiological) not normally present, or an unusually high concentration of a naturally occurring substance, in water, soil, or other environmental medium.

DRAWDOWN

The vertical distance groundwater elevation is lowered, due to the removal of groundwater. The distance between the static water level and the surface of the cone of depression.

FLOWING ARTESIAN WELL

When the top of a well in a confined aquifer is below the potentiometric surface, water will flow out of the well under pressure.

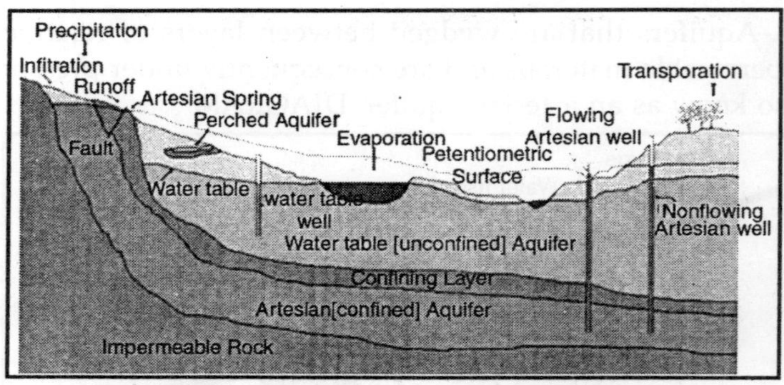

GROUNDWATER

Water occurring in the zone of saturation in an aquifer or soil. Water beneath the surface of the earth which saturates the pores and fractures of sand, gravel, and rock formations.

GROUNDWATER BARRIER

Rock or artificial material with a relatively low permeability that occurs (or is placed) below ground surface, where it impedes the movement of groundwater and thus may cause a pronounced difference in the heads on opposite sides of the barrier. Also called a confining layer.

HEAD

Height of the column of water at a given point in a groundwater system above a datum plane such as mean sea level. It is also a measure of pressure as in a confined aquifer where the elevation of the groundwater does not equal its head because it is under pressure.

HYDRAULIC CONDUCTIVITY (K)

A coefficient of proportionality describing the rate at which water can move through a permeable medium. It is a function of the porous medium and the fluid. The rate of flow of water in gallons per day through a cross-section of one square foot under a unit hydraulic gradient (gpd/ft^2), at the prevailing temperature.

HYDRAULIC CYCLE

The continuous circulation of water between the earth and the atmosphere, through condensation, precipitation, runoff, percolation, evaporation, transpiration, groundwater storage and seepage, and re-evaporation into the atmosphere.

HYDRAULIC GRADIENT (I)

Slope of a water table or potentiometric surface. Groundwater flows from points of high elevation and pressure to points of low elevation and pressure. The difference in hydraulic head divided by the distance along the flowpath.

IMPERMEABLE

Characteristic of geologic materials that limit their ability to transmit significant quantities of water under the head differences normally found in the subsurface environment.

LEAKING UNDERGROUND STORAGE TANK (LUST)

An underground tank which has a structural rupture and

its contents are leaving their containment and entering the surrounding environment.

NONPOINT SOURCE

Pollution of the water from numerous widespread locations that are hard to identify and pin-point. Ex. agri-chemicals through leaching or runoff or any conveyance not meeting the definition of point source.

PERCHED AQUIFER

An aquifer containing unconfined (unpressurized) groundwater held above a lower body of groundwater by an unsaturated zone; often a result of clay lenses in the soil strata.

PERMEABILITY

Capacity of a rock or soil material to transmit a fluid.

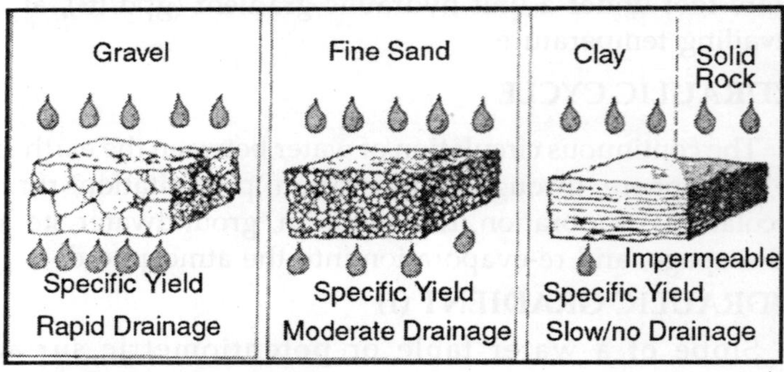

POINT SOURCE

Pollution of water from one place in a concentrated manner that is easy to identify. Ex. leaking underground storage tank, discharge pipe from a sewage treatment plant, any pipe, ditch, channel, tunnel, conduit, well, discrete fissure, container, rolling stock, animal feeding operation, or landfill, from which pollutants are or may be discharged.

POLLUTION PLUME

An area of a stream or aquifer containing degraded water resulting from migration of a pollutant. It extends from the source of contamination to another point in the direction of the water flow.

POROSITY

The capacity of soil or rock to hold water. The ratio of the volume of void spaces in a rock or sediment to the total volume of the rock or sediment.

POTABLE WATER

Suitable for human consumption as drinking water.

POTENTIOMETRIC SURFACE

An imaginary surface formed by measuring the level to which water will rise in wells of a particular aquifer. For an unconfined aquifer the potentiometric surface is the water table; for a confined aquifer it is the static level of water in the wells. (Also known as the piezometric surface.) DIAGRAM

PUBLIC WATER SYSTEM (PWS) (EPA)

Any water system providing water for human consumption for an average of at least 25 persons per day (or 15 or more service connections) and in use for at least 60 days each year. Further defined as follows:

Community Water System

A Public Water System which serves at least 15 services connections used by year-round residents or regularly serves at least 25 year-round residents.

Transient, Non-Community Water Systems

A Public Water System that is not a community water system. These systems do not regularly serve at least 25 of the same persons over six months per year.

Non-Transient, Non-Community Water System

A Public Water System that is not a community water system and that regularly serves at least 25 of the same persons over 6 months per year.

RECHARGE AREA

Area of land allowing water to pass through it into an aquifer by surface infiltration.

This process occurs naturally when rainfall filters down through the soil or rock into an aquifer, usually in the higher gradient section overlying the aquifer.

STATIC WATER LEVEL

The level of water in a well that is not being pumped. (The drawdown has been recharged by the surrounding groundwater.)

SATURATED ZONE

The portion of subsurface soil and rock where every available space is filled with water. Aquifers are located in this zone.

TIME OF TRAVEL (TOT)

The time required for a particle of water to move in the saturated zone from a specific point to a well.

A line can be drawn around the area for which groundwater is expected to reach the well within a chosen period.

TRANSMISSIVITY

A measure of the ability of an aquifer to transmit water. The rate at which water is transmitted through a unit width of an aquifer under a unit hydraulic gradient.

Transmissivity values are typically given in gallons per day through a vertical section of an aquifer one foot wide and extending the full saturated height of an aquifer under a hydraulic gradient of 1 (gpd/ft).

UNCONFINED AQUIFER

An aquifer which the water table is its upper boundary. Because the aquifer is not under pressure the water level in a well is the same as the water table outside the well. An unconfined aquifer is near to the earth's surface causing it to be easily recharged as well as contaminated.

UNDERGROUND STORAGE TANK (UST)

A tank system, including its piping, that has at least 10% of its volume underground

UNSATURATED ZONE

An area, usually between the land surface and the water table, where the openings or pores in the soil contain both air and water.

WATERSHED

All the land area and water within the confines of a drainage divide in which all surface runoff will drain through one point, such as a stream or river. Determined by topographic high points.

WATER TABLE

The water level of an unconfined aquifer, below which the pore spaces are saturated. The water table depth fluctuates with climate conditions on the land surface above and is usually gently curved and follows a subdued version of the land surface topography.

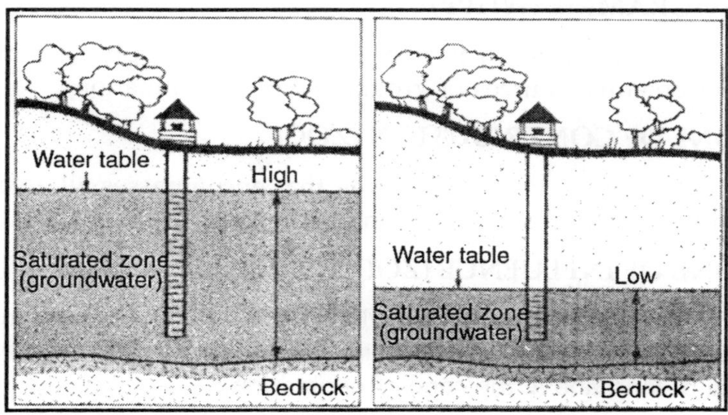

WELL

An opening in the surface of the earth for the purpose of removing fresh water.

WELL FIELD

An area containing two or more wells supplying a public water supply system.

WELLHEAD

The physical structure, facility, or device at the land surface from or through which groundwater flows or is pumped from subsurface, water-bearing formations.

WELLHEAD PROTECTION AREA (WHPA)

A geographically designated surface and subsurface area surrounding a water well or wellfield, supplying a Public Water System, through which contaminants are reasonably likely to move toward and reach such water well or well field.

This is an area where groundwater protection is emphasized. A WHPA includes at least the delineated capture zone area, but could be influenced by other factors such as existing land use and zoning, site and facility identification and location, political boundaries, natural features, and environmentally sensitive areas.

WELLHEAD PROTECTION PROGRAMME (WHPP)

A programme to protect public water supply systems from potential sources of groundwater contamination.

ZONE OF CONTRIBUTION (ZOC)

The area surrounding a well that encompasses all areas and features that supply groundwater to the well.

ZONE OF INFLUENCE (ZOI)

The area of groundwater which is affected by the pumping of a well. The faster the pumping rate the larger the area. The area of land above the cone of depression.

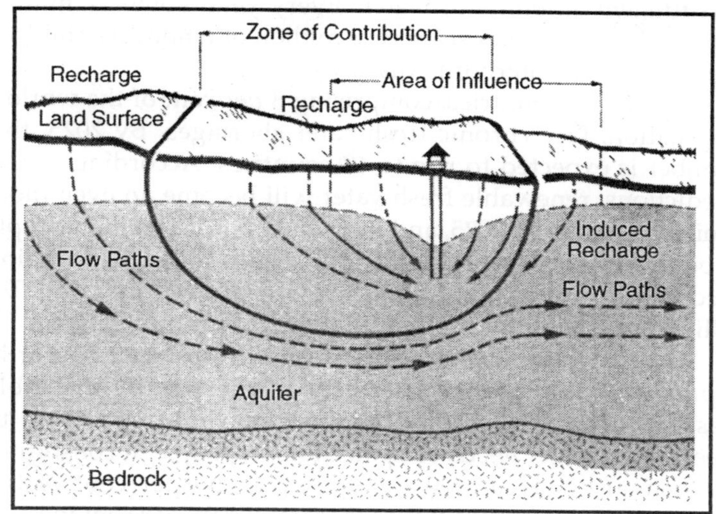

ZONE OF TRANSPORT (ZOT)

The area that contains the groundwater that will reach the well within a specified time. Similar to time of travel.

ARTESIAN WELLS

A closed artesian aquifer is confined by an overlying impermeable body of rock, which prevents any water from filtering down into the aquifer. Instead, water enters the tilted aquifer layer through a recharge area, where the aquifer rock is exposed at higher elevations.

The flow in an artesian aquifer resembles water flowing through a J-shaped tube. Water added on the tube's long side provides enough pressure to drive the water upward on the tube's shorter side.

The above verse describes how Prophet Musa's (as) people asked him for water and how he provided places where each tribe could drink. Clearly, his people were suffering from a shortage of water. Such shortages still exist, for more than 1 billion people today lack access to clean water, and 2.4 billion still live without improved sanitation. According to projected estimates, by 2025 about 5 billion people will not have access to sufficient amounts of water. Every year, some 12 million people die from water scarcity; 3 million of whom are children who die from waterborne diseases.

Today, 31 countries, comprising 8 per cent of the world's population, face chronic freshwater shortages. By 2025, this number is expected to rise to 48 countries. According to UN predictions, renewable freshwater will become an even more limited resource by 2025, and the number of 131 million people experiencing water problems will rise to either 817 million (according to low population growth projections) or 1.079 billion (according to high population growth projections).

Groundwater, the largest source of fresh water on Earth, represents more than 90 per cent of the readily available freshwater reserves and is therefore of vital importance to meeting the water needs of up to 2 billion people. It constitutes the primary source of water for up to 50 per cent of the American population, a figure that rises to 95 per cent in rural areas. Groundwater is also the safest and most reliable source of fresh water. At the same time, this water can be used to produce geothermal energy and save energy by using heat pumps.

When the water sucked up from the soil meets an impermeable underground layer, it collects there and forms a water source. This water is then brought to the surface by the artesian method. Artesian springs are formed by sedimentary rocks that can store underground water.

The fact that artesian wells are drilled in rocky areas runs parallel to the description in the Qur'an. Given that Allah commanded Prophet Musa (as) to strike the rock, Surat al-A'raf 160 may be indicating this method. (Allah knows best.) The verb idrib, translated as "strike," can also mean "to raise, to open." Thus, this verse may be describing a water source being opened by the raising of the rock.

As a result, pressurized water may have emerged, as described in the verb inbajasat (to pour out, flow freely, bubble up, flow), just as happens with artesian wells. If sufficient pressure forms, water can continue to flow to the surface without the need for a pump.

Allah is He Who created the heavens and the earth and sends down water from the sky and by it brings forth fruits as provision for you

It is particularly striking that current solutions for dealing with water scarcity use underground water resources. In fact, one of the most effective methods of doing so is the artesian well. In other words, we might be copying Prophet Musa's (as) example of striking or lifting the rock without even knowing it. Surat al-A'raf 160 may therefore be a reference to artesian wells, the first of which was opened in 1126 in the French region of Artois

Artesian wells tap the confined groundwater trapped below impermeable strata beneath the surface of the earth. Natural pressure forces the water to rise, sometimes above the surface of the earth (UNDERGROUND WATER). Although extensive drilling for artesian wells did not begin until the 1880s, in 1857 and 1858 John Pope experimented with artesian wells on the plains of West Texas, and in 1858 a bill introduced byForbes Britton became a law authorizing the drilling of public artesian wells on the road between San Antonio and Laredo.

These early experiments were typical of the many mistakes made, through lack of knowledge, in an effort to secure water in the arid portion of the state. The artesian-well area is largely confined to the Coastal Plain east of the Balcones Escarpment, although there have been other isolated artesian areas, most notable of which were located at El Paso, Fort Stockton, and Balmorhea. By 1897 there were 458 flowing and 506 nonflowing wells in the Black and Grand Prairie regions, somewhat fewer on the Coastal Plain, and none on the High Plains.

Around the turn of the century there were six artesian districts in Texas: Coastal Prairie system, Hallettsville system, Carrizo system, Black and Grand prairies system, Trans-Pecos Basin system, and Stevens County and Jack County systems. At that time the city of Galveston received its water from approximately thirty-three artesian wells, and Houston had about 100.

In 1905 the area comprising most of South Texas from twenty miles south of Corpus Christi to the Rio Grande valley was referred to as the Artesian Belt.

Farmers were just beginning to realise the agricultural potential of South Texas and relied on irrigation from artesian wells to provide water to crops.

In a 1905 edition of*Farm and Ranch* magazine one article stated that 200 wells were in operation, ranging in depths from 600 to 800 feet and averaging flows from 200 to 600 gallons per minute.

Wells were also touted in promotional brochures, such as the ad for the town of Sarita, Texas, which estimated that an artesian well would irrigate 100 to 200 acres. Costs for digging those wells ranged from $1,000 to $1,500.

Water from artesian wells was included in the general irrigation act of 1913 (WATER LAW), and in May 1931 the legislature passed a law to prevent the waste of artesian water, which had become an important source of water supply for numerous cities and irrigation projects. Throughout the later part of the twentieth century more cities looked to other options for municipal water such as the use of surface reservoirs.

Though the use of wells remained the property of the individual owner, increased government regulation such as the issuing of permits and possibilities of restrictions became the focus of debate. According to Texas law a person can drill an artesian well for domestic or stock purposes, but the well should be properly cased. Records of depth, thickness, and the types of strata penetrated should be kept. Reports of new artesian wells are made to the Texas Water Commission. Those using wells for other than domestic needs must submit an annual report regarding the well level, quantities used, and methods of use.

PROVINCES AND ITS ROLE

GEOLOGICAL HAZARD

A geologic hazard is a natural geologic event that can

endanger human lives and threaten human property. Earthquakes, geomagnetic storms, landslides, sinkholes, tsunamis, and volcanoes are all types of geologic hazards. The U.S. Geological Survey (USGS) provides real-time hazard information on earthquakes, landslides, geomagnetics, and volcanoes, as well as background information on all the types of hazards described below.

Earthquakes

Fig. More than 6 Feet was Added to this Fault Scarp by Vertical Movement in the 1983 Borah Peak

The term "earthquake" refers to the vibration of the Earth's surface caused by movement along a fault, by a volcanic eruption, or even by manmade explosions. The vibration can be violent and cause widespread damage and injury, or may be barely felt.

Most destructive earthquakes are caused by movements along faults. Earthquakes can occur at the surface of the Earth or as deep as 400 miles below the surface. An earthquake can trigger additional hazards such as landslides or tsunamis.

Earthquakes occur all over the world and often occur without significant warning. These geohazards can have far-reaching affects on humans and on the surface of the Earth. Small, localized earthquakes may cause no noticeable damage and may not even be felt by people living in the affected area. In contrast, a large earthquake may cause destruction over a wide area and be felt thousands of miles away.

Fig. Subsidence at Government School

The Earth is formed of several layers that have very different physical and chemical properties. The outer layer, which averages about 43 miles (70 kilometres) in thickness, consists of about a dozen large, irregularly shaped plates that slide over, under, and past each other on top of the partly molten inner layer.

The plate boundaries are fault zones, and are where most earthquakes occur. In fact, the locations of earthquakes and the kinds of ruptures they produce help scientists define the plate boundaries. There are three types of plate boundaries: spreading zones, transform faults, and subduction zones. At spreading zones, molten rock rises, pushing two plates apart and adding new material at their edges. Most spreading zones are found in oceans; for example, the North American and Eurasian plates are spreading apart along the mid-Atlantic ridge. Spreading zones usually have earthquakes at shallow depths (within 19 miles (30 kilometres) of the surface).

Earthquakes can also occur within plates, although plate-boundary earthquakes are much more common. Less than 10 per cent of all earthquakes occur within plate interiors. As plates continue to move and plate boundaries change over geologic time, weakened boundary regions become part of the interiors of the plates. These zones of weakness within the continents can cause earthquakes in response to stresses that originate at the edges of the plate or in the deeper crust. The New Madrid earthquakes of 1811-1812 and the1886 Charleston earthquake occurred within the North American plate. Earthquake strength is measured as both magnitude and intensity. Magnitude measures the relative strength of an earthquake and is recorded with the Richter scale. Each earthquake only has one magnitude. People usually cannot feel earthquakes with magnitudes of 3.0 or less. Intensity measures the severity of an earthquake in terms of its effect on humans, structures, and the land surface. The USGS usually uses the Modified Mercalli intensity scale to describe earthquake intensity. The intensity of a given earthquake will vary from place to place.

We tend to picture most earthquake damage as resulting directly from ground shaking, but there are many other related impacts from an earthquake. For example, ground shaking can result in soil liquefaction, damage to dams or levees with resultant flooding, landslides, and fires caused by ruptured fuel and power lines. In addition, earthquakes may trigger tsunamis or seiches. Structural damage or collapse may be caused by any of these effects, which may be local or may occur hundreds or even thousands of miles from the epicentre of the earthquake. A Federal Emergency Management Agency study considered just capital (damages to buildings and their contents) and income-related costs, and provided an estimate of $4.4 billion as the minimum average annualized loss due to earthquakes in the United States.

Most earthquakes in the United States occur in Alaska and California, although Hawaii, Nevada, Washington, and Idaho also experience many earthquakes. Four of the five largest earthquakes in the United States occurred in Alaska. The USGS Earthquake Hazards Programme maintains detailed records on historical earthquakes and continuously monitors earthquake activity around the world.

While we can't prevent earthquakes or even accurately predict when they will occur, we can take steps to lessen their human impact. In the United States and many other countries, building codes take into account the local earthquake risk so that buildings and other structures can be designed to withstand all but the most severe earthquakes. In addition, those who live in earthquake-prone areas should know how to be prepared for an earthquake and what to do if one occurs.

GEOMAGNETIC STORMS

The Earth has an associated magnetic field, referred to as the geomagnetic field, which is caused by electric currents both within the Earth and in the area surrounding the Earth. The geomagnetic field is what causes a compass to point north. The interaction of the geomagnetic field with the solar wind is what gives us aurora - the Northern or Southern Lights.

Fig. Electric Power line Towers.

Fig. Line Towers

Why is geomagnetism considered a geologic hazard? Occasionally the earth experiences a "magnetic storm", which is a rapid variation in the geomagnetic field, caused either by a gust in the solar wind or by a temporary linking of the Sun's magnetic field with the geomagnetic field. Magnetic storms can disturb long-range radio communication, degrade global positioning systems, damage satellites, affect long-distance pipelines, and produce surges on electric power grids resulting in blackouts. In addition, magnetic storms can expose astronauts and high-altitude pilots to increased levels of radiation. Large magnetic storms can produce dramatic auroral displays. Variations in the magnetic field don't have a direct impact on human health, but do affect the technology that is important to our modern society.

Because the geomagnetic field covers the entire Earth, problems caused by geomagnetic storms can occur almost anywhere. The National Oceanic and Atmospheric Administration (NOAA), National Weather Service, Space Environment Centre (SEC) monitors geomagnetic storm activity and provides real-time information on their Space Weather Now site. The SEC has defined five types of solar radiation storms, ranging from mild to extreme; their definitions include a description of the possible damaging effects of each class of magnetic storm.

The USGS National Geomagnetism Programme provides additional information to the public on the geomagnetic field and geomagnetic hazards.

LANDSLIDES

A landslide is the movement of soil, rock, or other earth materials, downhill in response to gravity. Landslides include rock falls and topples, debris flows and debris avalanches, earth flows, mudflows, creep, and lateral spread of rock or soil.

Frequently landslides occur in areas where the soil is saturated from heavy rains or snowmelt. They can also be started by earthquakes, volcanic activity, changes in groundwater, disturbance or change of a slope by man-made construction activities, or any combination of these factors. A variety of other natural causes may also result in landslides, and they may trigger additional hazards, such as tsunamis caused by submarine landslides. A landslide occurs when the force that is pulling the slope downward (gravity) exceeds the strength of the earth materials that compose the slope.

Rock falls or topples are usually sudden and occur on steep slopes. In a rock fall, rocks may fall, bounce, or roll down the slope. A topple occurs when part of a steep slope breaks loose and rotates forward.

A debris flow is a combination of water-saturated loose soil, rock, organic matter, and air, with material varying in size from grains of clay to large boulders. Such flows are formed when loose masses of unconsolidated wet debris become unstable. A lahar is a special type of debris flow that originates from the slopes of a volcano

Water for a debris flow may be supplied by rainfall, by melting of snow or ice, or by overflow of a lake, and the flow may be either hot or cold, depending on how it starts and the temperature of the constituent debris.

When moving, a debris flow resembles a mass of wet concrete and tends to flow along channels or stream valleys. It can travel down a hillside at speeds up to 200 miles per hour (more commonly, 30 to 50 miles per hour), depending on the slope angle, the water content, and the type of earth and debris in the flow. Burned areas are particularly susceptible to debris flows. Very rapidly moving debris flows are known as debris avalanches.

Moderate slopes, and consist of saturated soil or fine-grained rock deposits that flow downhill. Dry earthflows are also possible. A mudflow is an earthflow consisting of material that is wet enough to flow rapidly.

Creep is the imperceptibly slow, steady, downward movement of slope-forming soil or rock. Creep can occur seasonally, where movement is within the depth of soil affected by seasonal changes in soil moisture and soil temperature, or can be continuous or progressive. Creep is indicated by curved tree trunks, bent fences or retaining walls, tilted poles or fences, and small soil ripples or ridges.

Most landslides happen on steep or moderate slopes, but lateral spreads usually occur on very gentle slopes or in flat terrain.

These spreads are caused by liquefaction, the process whereby saturated, loose sediments that will not stick together (usually sands and silts) are transformed from a solid into a liquefied state. Lateral spread is usually triggered by rapid ground motion, such as that experienced during an earthquake, but can also be artificially induced.

The combination of two or more types of landslides is known as a complex landslide. Landslides constitute a major geologic hazard because they are widespread, occurring in all 50 States, and because they cause more than $2 billion in damages and more than 25 fatalities on average each year. Casualties in the United States are primarily caused by rockfalls, rock slides, and debris flows.

Worldwide, landslides cause. Earthflows usually occur on thousands of casualties and billions in monetary losses annually. The USGS Landslide Hazards Programme collects and distributes information on landslides to the public, scientists, and civil authorities, and works to reduce losses and deaths from landslides.

One of the largest landslides in the world the 20th century occurred at Mount St. Helens, Washington, in 1980. A moderate earthquake caused roughly 1.7 cubic miles of rocks and mud to break free and slide down the side of the volcano, releasing pent-up pressure to produce the major eruption of May 18. Although this was the largest landslide recorded in historic time, fewer than 60 people were killed because most residents and visitors had been evacuated. The most costly landslide in U.S. history was a relatively slow-moving event in Thistle, Utah, in the spring of 1983. The landslide, caused by the wet El Nino winter of 1982-83, dammed the Spanish Fork River and buried U.S. Highway 6 and the main line of the Denver and Rio Grande Western Railroad. The town of Thistle was inundated under the floodwaters rising behind the landslide dam. Total losses were estimated at more than $400 million in 1983 dollars.

SINKHOLES

Sinkholes, like landslides, are a form of ground movement that can happen suddenly and with little warning, and that can cause major damage. Sinkholes are common where the rock below the land surface is limestone, carbonate rock, salt beds, or rocks that can naturally be dissolved by ground water circulating through them.

As the rock dissolves, spaces and caverns develop underground. Sinkholes are dramatic because the land usually stays intact for a while until the underground spaces just get too big, then a sudden collapse occurs. These collapses can be small and have little impact on people, or they can be huge and can occur where a house, road, or other structure is on top.

In the United States, the most damage from sinkholes tends to occur in Florida, Texas, Alabama, Missouri, Kentucky, Tennessee, and Pennsylvania, in what is known as karst topography. Sinkholes are also a problem in many other places around the world.

TSUNAMIS

Tsunamis are large, destructive waves that are caused by the sudden movement of a large area of the sea floor. Tsunamis are often incorrectly called tidal waves, but unlike regular ocean tides they are not caused by the gravitational pull of the moon and sun.

Most tsunamis are caused by earthquakes, some are caused by submarine landslides, a few are caused by submarine volcanic eruptions and on rare occasions they are caused by a large meteorite impact in the ocean.

The December 26, 2004 magnitude 9.0 earthquake near Sumatra produced the largest trans-oceanic tsunami in over 40 years, and killed more people than any tsunami in recorded history.

The Krakatau volcanic eruption of 1883 generated giant waves reaching heights of 125 feet above sea level, killing thousands of people and wiping out numerous coastal villages.

While tsunami means "harbor wave" in Japanese, a tsunami is actually a series of large waves created by the sudden movement of the seafloor. The energy generated by the earthquake or other event is transmitted through the water as a large train of waves, but the movement of these waves is very different from the movement of waves generated by wind. NASA's Physics Behind the Wave explains the structure of tsunamis. Tsunamis can travel rapidly across oceans, causing destruction far from the location where they were generated.

All oceanic regions of the world experience tsunamis, although tsunamis in the Atlantic, Mediterranean, and Caribbean tend to be smaller and less destructive than those in the Pacific and Indian Oceans. About 90 per cent of recorded tsunamis occur in the Pacific Ocean.

The reasons for this lie in the geologic structure of the Pacific basin - the ocean is surrounded by a geologically active series of mountain chains, deep ocean, trenches, and island arcs, sometimes called "the ring of fire." The earthquakes and volcanic eruptions that occur in the ring of fire are the source of many tsunamis.

NavyThe height of a tsunami in the deep ocean is small - usually about 1 foot - and they cannot be seen or felt by ships at sea. The distance between wave crests can be more than 100 miles. The speed at which the tsunami travels decreases as water depth decreases.

In the deep waters of the mid-Pacific, a tsunami can reach a speed of more than 500 miles per hour, but in the shallow waters near land the speed drops to 100 miles per hour or less. As tsunamis reach shallow water the height of the waves increases dramatically, and can reach 100 feet or more.

These huge waves can wash far inland, carrying large amounts of debris, destroying buildings and other structures, causing widespread flooding, and dramatically altering shorelines. Most tsunamis consist of a series of waves, and the first wave to reach shore may not be the largest.

Locally generated tsunamis may reach a shoreline with only a few minutes warning, while distant events may allow several hours warning. Warning signs of an approaching tsunami include a strong earthquake felt near the shore or a rapid fall in the water level - like a sudden and extremely low tide. Either of these signs should be taken as a warning that a tsunami is imminent and that coastal areas should be immediately evacuated.

In addition, many coastal areas have tsunami alert systems that sound sirens or provide information through local media. The United States has a tsunami warning system in place for the west coast, Hawaii, and Alaska. The West Coast and Alaska Tsunami Warning Centre (WCATWC) in Palmer, Alaska, provides information for Alaska, Washington, Oregon,

California, and British Columbia. WCATWC also provides online tsunami safety advice. Information for the remaining portions of the Pacific basin is supplied by the Pacific Tsunami Warning Centre in Hawaii.

The tsunami warning centres issue two types of bulletins to advise of a possible approaching tsunami. A Tsunami Watch Bulletin is released when an earthquake occurs with a magnitude of 6.75 or greater on the Richter scale.

A Tsunami Warning Bulletin is released when information from tidal stations indicates that a potentially destructive tsunami exists. Tidal stations record information about the water around them and issue a warning when characteristics of the sea begin to match those of a potential tsunami.

While we can't prevent tsunamis, we can take steps to lessen their impact. Those who live in or visit tsunami-prone areas should know the warning signs of an approaching tsunami, and what to do when a tsunami is imminent.

VOLCANOES

A volcano is a vent at the Earth's surface through which magma and associated gases erupt, and also the cone built by eruptions.

A volcano that is currently erupting or showing signs of unrest (earthquakes, gas emissions) is considered active. A volcano that is not currently active but which could become active again is considered dormant. Extinct volcanoes are those considered unlikely to erupt again.

Volcanic eruptions are one of Earth's most dramatic and

violent agents of change. They pose significant geologic hazards because their eruptions and associated activities can affect large areas and go on for extended periods of time.

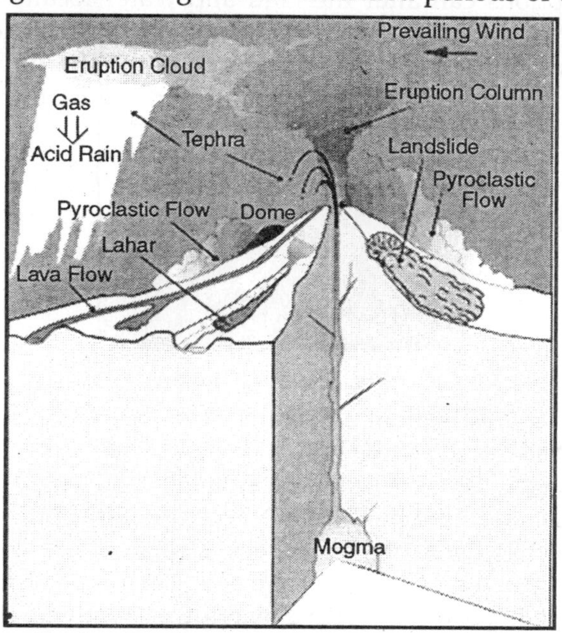

Many kinds of volcanic activity can endanger the lives of people and property, and the affects of these activities are felt both close to and far away from the volcano. Explosive eruptions can spread lava, gas and other materials over a wide area, and may drastically alter the landscape. Slow eruptions or flows can also alter landscapes, while associated earthquakes, atmospheric effects, landslides, and floods all may damage or destroy property and threaten human lives.

Some volcanic eruptions are mild and slow, while others are powerful and dramatic. An eruption happens when magma, gases, or steam break through vents in the Earth's surface. A mild eruption may simply discharge steam and other gases, or quietly extrude lava.

A strong eruption can consist of violent explosions that send great clouds of gas-laden debris into the atmosphere, or may consist of explosions that blast sideways from a collapsed portion of the volcano, as happened in the 1980 eruption of Mount St. Helens.

Eruptions can alter the land and water locally through lava flows, lahars, pyroclastic flows, and landslides. An eruption cloud of ash and gas may spread the impact of a volcano over many miles or even around the Earth.

Fig. Flow of a Lava

A lava flow is molten rock that has reached the surface of the Earth. It may flow quickly or slowly, but destroys everything in its path, including vegetation and manmade structures, and may bury homes and agricultural land under tens of feet of hardened black rock. People are rarely able to use land buried by lava flows or to sell it for more than a small fraction of its previous worth. Lava flows usually do not travel far from their volcanic source.

Lava entering the sea poses special risks. With temperatures higher than 2,000 degrees Fahrenheit (1,100 degrees Celsius), lava can instantly transform seawater to steam, causing explosions that blast hot rocks, water, and molten lava fragments into the air. A lava delta, created as lava enters the sea, looks like a stable platform that extends tens to hundreds of feet into the ocean. However, the lava delta may not be well supported and can collapse into the ocean with little or no warning.

A lahar is a mixture of volcanic ash, rock, debris, and water that can travel quickly down the slopes of a volcano. They are generated when a high volume of hot or cold water mixes with ash and rock and starts down slope. The water may come from melting snow or ice, heavy rainfall during an eruption, or the breakout of a lake. When moving, a lahar looks like a mass of wet concrete. As a lahar rushes downstream from a volcano, its size, speed, and the amount of water and rock debris it carries constantly change. The beginning surge of water and rock debris often erodes rocks and vegetation from the side of a volcano and along the river valley it enters. This initial flow can also incorporate water from melting snow and ice or from the river it overruns. By eroding rock debris and incorporating additional water, lahars can easily grow to more than 10 times their initial size. But as a lahar moves farther away from a volcano, it will eventually begin to lose its heavy load of sediment and decrease in size.

A pyroclastic flow is a rapidly-moving mixture of hot, dry rock fragments, ash, and hot gases which knocks down, buries, or burns everything in its path. Pyroclastic flows are caused by explosive eruptions or by the collapse of a lava flow, can reach temperatures as high as 1,300 degrees Fahrenheit (700 degrees Celsius), and may melt snow and ice to cause lahars. These flows vary considerably in size and speed, but even relatively small flows can destroy buildings, forests, and farmland. Even on the margins of pyroclastic flows, death and serious injury to people and animals may result from burns and inhalation of hot ash and gases.

Volcanic landslides are common and can be caused by an eruption or associated heavy rainfall, by an earthquake under the volcano, or by the collapse of a slope weakened by underlying volcanic activity. A landslide caused by collapse of part of the volcano's cone may also trigger an eruption as pressure on the underlying volcanic systems is decreased. Historically, landslides have caused explosive eruptions, buried river valleys with tens to hundreds of feet of rock debris, generated lahars, triggered waves and tsunami, and created deep horseshoe-shaped craters. Moving rapidly and with great momentum, a large volcanic landslide may flow up and over ridges, and may cause damage far from the volcano.

Tephra is fragments of volcanic rock and lava that are blasted into the air by explosions or carried upward by hot gases in eruption columns or lava fountains. These fragments may be as small as ash or as large as several feet in diametre. Tephra includes combinations of pumice, glass shards, crystals from different types of minerals, and shattered rocks. Large tephra typically falls back to the ground near the volcano while smaller fragments are carried away by wind. Volcanic ash, the smallest tephra fragments, can travel hundreds to thousands of miles downwind from a volcano. Ash usually covers a much larger area and disrupts the lives of far more people than the other more lethal types of volcano hazards. Ash fall may injure livestock and crops, collapse buildings, damage communications and power-supply facilities, cause driving and visibility problems, damage or disable aircraft, and cause respiratory and eye irritation problems in people.

SurveyMagma contains dissolved gases that are released into the atmosphere during eruptions, primarily as acid aerosols (tiny acid droplets), compounds attached to tephra particles, and microscopic salt particles. Volcanic gases may also escape continuously into the atmosphere from the soil, volcanic vents, fumaroles, and hydrothermal systems. The volcanic gases that pose the greatest potential hazard to people, animals, agriculture, and property are sulfur dioxide, carbon dioxide, and hydrogen fluoride. Sulfur dioxide gas can lead to acid rain and air pollution downwind from a volcano, and large amounts may lead to lower surface temperatures and promote depletion of the Earth's ozone layer. Concentrations of carbon dioxide gas can be lethal to people, animals, and vegetation, while hydrogen fluoride can contribute to acid rain and is a powerful irritant that can deform or kill animals.

Scientists monitor active volcanoes and try to anticipate when an eruption will occur. Volcano monitoring methods detect and measure changes in the state of a volcano caused by magma movement beneath the volcano. Rising magma typically will trigger numerous earthquakes, cause swelling or subsidence of a volcano's summit or flanks, and lead to the release of volcanic gases from the ground and vents.

In the United States, the USGS Volcano Hazards Programme has established a series of volcano warning schemes that are used to notify the public and civil authorities of impending volcanic activity orVolcanic activity since 1700 has killed more than 260,000 people, destroyed entire cities and forests, and severely disrupted local economies for months to years. Even with our improved ability to identify hazardous areas and warn of impending eruptions, increasing numbers of people face certain danger. Scientists face a formidable challenge in providing reliable and timely warnings of eruptions to so many people at risk.

Chapter 8

Building Stones Engineering

PROPERTIES OF ROCKS

The seismic properties of rocks are usually thought of in terms of P-wave velocity, V_p, and S-wave velocity, V_s, which are both functions of density (seismic waves, principles). A knowledge of V_p, V_s, and density can tell us a great deal about the composition and physical state of materials in the earth. The Earth is, however, much more complex than this simple approach would indicate. Laboratory studies on individual mineral crystals are capable of great detail and precision.

The situation is very different for measurements based on the analysis of body and surface waves travelling through the Earth. It has, for example, commonly been the practice to assume that materials are mechanically isotropic so that only two elastic constants are required.

The constants measured are usually V_p and V_s. Modern studies are often able to go beyond this simple assumption but usually stop at determining V_p, V_s, density, Q, the quality factor (assumed to be a constant), and anisotropy expressed as a per centage. As a practical matter, this amounts to measuring from three to five elastic constants out of a possible 21, and recognizing that observable differences in velocity exist which are a function of the direction of travel.

Seismologists desire to measure physical properties to the greatest level of detail possible. Their aim is to determine not only the composition of the Earth but also its physical state. The question of composition is complex because rocks are combinations of minerals which each have their own seismic properties. In general, the velocities of the crystalline rocks which constitute most of the Earth's lithosphere are inversely proportional to the silicon dioxide content and directly proportional to the content of iron and magnesium oxides.

Many factors, such as porosity, pressure, and temperature, affect the seismic velocity of a particular rock. Figure, which is a compilation of typical values, illustrates some relationships which are generally true. Unconsolidated sediments can have very low velocities. Clastic sedimentary rocks (sandstone and shale) are 'slower' than carbonate sedimentary rocks (limestone and dolomite). For typical crystalline rocks, granite (felsic) is slower than gabbro (mafic), which is slower than peridotite (ultramafic).

These relationships are generally reflected in the velocity structure of the Earth's lithosphere, in which the granitic upper crust generally has higher velocities than the sedimentary rocks which cover it but has lower velocities than the more mafic lower crust. The ultramafic mantle, which is very rich in iron, has even higher velocities.

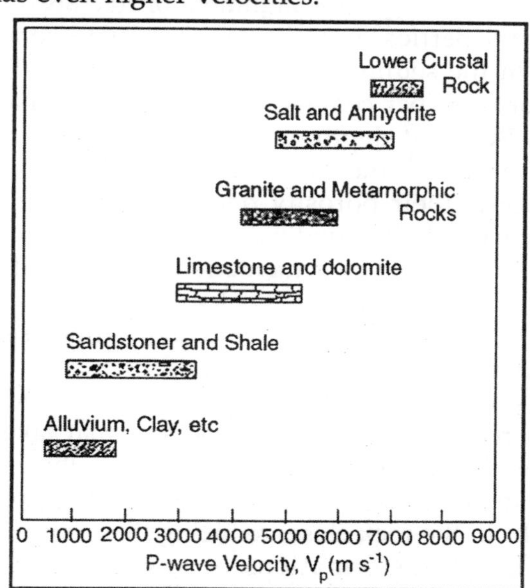

Fig. P-wave Velocities (V_p) for Various Rock Types.

Such factors as the degree of alignment of particular minerals govern properties such as the level of anisotropy. For example, an early large-scale observation of anisotropy was provided by P-wave velocities measured in the mantle near ocean spreading centres. Here, alignment of the highly anisotropic mineral olivine is sufficient to cause velocities measured from refraction profiles that are recorded in the direction of spreading to be higher than those for profiles recorded parallel to the spreading centre.

This alignment is particularly interesting because it is related to the flow which reflects plate movements. As a result there is much interest in measuring anisotropy elsewhere in the mantle. Some metamorphic rocks and shales have mineral alignments which are clear in hand specimens and which are reflected in significant anisotropy, but when viewed on a larger scale most crustal rocks do not appear to be very anisotropic. However, detailed velocity measurements, such as those made by logging devices which are lowered down boreholes, often reveal significant anisotropy in individual rock units, which is sometimes a result of the alignment of fractures.

Other properties of rocks can dominate composition, especially in the sedimentary rocks found near the Earth's surface. Porosity (per centage of pore space) is particularly important. Burial at depth, giving rise to high overburden pressures, causes sedimentary rocks in particular to compact. In the process, their porosity decreases and their seismic velocity increases correspondingly. For example, pressure due to burial, accompanied by cementation, causes sand and mud to evolve into sedimentary rocks (sandstone and shale respectively), and their velocity usually increases as this evolution proceeds.

Increasing age also causes the velocity of sedimentary rocks to increase because many processes that take place in the Earth tend to decrease the porosity. Crystalline rocks also exhibit an increase in velocity with increasing pressure, as micro-cracks in the rock close, reducing their porosity. In the mantle, phase transformations of olivine result in more compact crystal structures which are believed to cause the seismic discontinuities that are consistently observed at several depths on a global basis. Temperature is another factor that affects the seismic properties of rocks. If other factors are held constant, an increase in temperature generally causes a decrease in seismic velocity. Since temperature increases with depth in the Earth, we might expect seismic velocity to decrease with depth. Except in small regions, this is not, however, the case because in general the temperature effect is offset by the effects of pressure and composition.

In many cases instances it is important to know V_p, V_s, and density (r) in order to study some phenomenon; for example, the variation in the amplitude of a wave reflected from a discontinuity in the Earth with the angle at which the wave is approaching. We may, however, be able to measure only V_p. Since we have good values for Poisson's ratio (s) for typical rock types, we can calculate V_s from a standard equation.

This relationship becomes very useful when we have independent information on density from gravity measurements or logging devices which measure the density of rocks penetrated by a borehole. In addition to Poisson's ratio, there is a good correlation between V_p and density. This relationship is based mostly on measuring the density and V_p for many different rock types. It has been presented in several forms.

These relationships are useful, but only approximate. One should not lose sight of the fact that many types of rocks share the same P-wave velocity, V_p (Fig. 1). More independent measurements of other seismic properties are thus highly desirable so that the type of rock and its physical state can be better determined.

The attenuation of seismic waves as they travel through the Earth is a complex subject which is the target of much research. In general, we find that the amplitude of a seismic wave dies off in proportion to the factor $e^{''ax}$, where x is the distance travelled and a is the attenuation coefficient. The attenuation coefficient is a function of frequency, and it becomes larger as the frequency increases. This preferential attenuation of higher frequencies is a fundamental limitation in seismology that restricts our ability to resolve (i.e. see) deep structures because we need the short wavelengths associated with high frequencies (seismic waves, principles). It would be a major task to tabulate *a* for many materials as a function of frequency, so it is fortunate that *a* is an approximately linear function of frequency for typical seismic waves. This relationship leads to the quality factor, *Q*, being a constant. The quality factor is thus a convenient measure of the degree to which a material absorbs seismic energy; high values of *Q* mean that attenuation is low. An early large-scale observation was that the cold down-going slabs in subduction zones had high *Q* values while the hot volcanic arcs above the subduction zones had low *Q* values.

ALKALI AGGREGATE REACTION

Alkali-aggregate reaction is the expansive reaction that takes place in PCC between alkali (contained in the cement paste) and elements within an aggregate. The most common is an alkali-silica reaction. This reaction, which occurs to some extent in most PCC, can result in map or pattern cracking surface popouts and spalling if it is severe enough.

The mechanism for this alkali-silica reaction proposed by Diamond is as follows:

Fig. Map/Pattern Cracking Resulting from an Alkali-Aggregate Reaction

ALKALINE DEPOLYMERIZATION

Cement (a high-alkali substance) can increase the solubility of non-crystalline silica and the rate at which it dissolves. Additionally, the cement will raise the pH of the surrounding medium which will affect the crystalline silica.

HYDROUS ALKALI SILICATE GEL

The initial dissolution of reactive silica then opens up the aggregate pore structure and allows more silica to dissolve into solution. The end result is alkali-silica gel that is formed in place. This gel formation is not expansive itself but it does destroy the integrity of the aggregate particle.

ATTRACTION OF WATER BY THE GEL

The gel attracts considerable amounts of water and expands. If the expansion is great enough, the resulting stress will crack the now-weakened aggregate and surrounding cement paste.

FORMATION OF A GEL COLLOID

After the gel ingests enough water, the water takes over and the substance becomes an alkali-silica gel disbursed in a water fluid. This fluid then escapes to surrounding cracks and voids and may partake in secondary reactions.

AVOIDING SUSCEPTIBLE AGGREGATES

Local experience may show that certain types of rock contain reactive silica. Typically rock types that may be susceptible are: siliceous limestone, chert, shale, volcanic glass, synthetic glass, sandstone, opaline rocks and quartzite. River rock is also typically susceptible.

POZZOLANIC ADMIXTURE

By reacting with the calcium hydroxide in the cement paste, a pozzolan can lower the pH of the pore solution. Additionally, the silica contained in a pozzolan may react with the alkali in the cement. This reaction is not harmful because it essentially skips the expansive water attraction step.

LOW-ALKALI CEMENT

Less alkali available for reaction will limit gel formation.

LOW WATER-CEMENT RATIO

The lower the water-cement ratio, the less permeable the concrete. Low permeability will help limit the supply of water to the alkali-silica gel. In sum, alkali-silica reactions are expansive in nature and occur in most PCC. If the reaction is severe enough it can fracture aggregates and surrounding paste resulting in cracking, popouts and spalling. There are several ways of avoiding this reaction, the simplest of which is just avoiding susceptible aggregate.

GROUTING

Features that can influence grouting design and construction include:

SPACING OF JOINTS

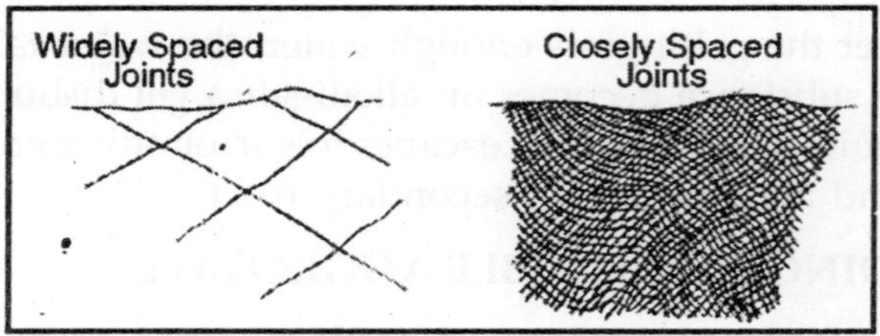

As far as cement grouting is concerned, it is the open, groutable joints that are of interest. If they are widely spaced, the grouting is usually easier than if closely spaced where troubles such as frequent surface leaks, collapsing holes and patchy penetrations can happen. These make for more expensive grouting, perhaps requiring special surface treatment.

JOINT WIDTHS AND CONTINUITY

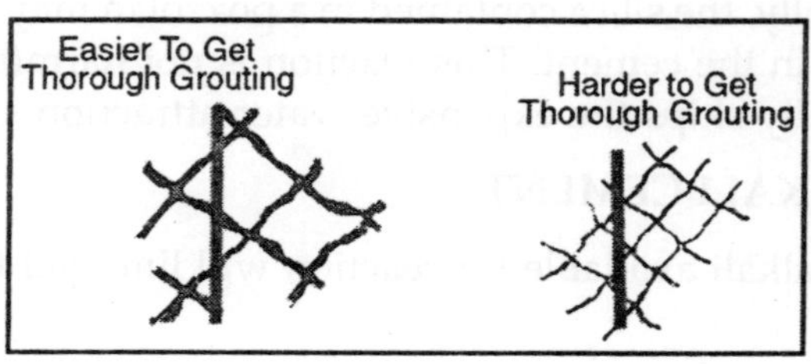

The easiest joints to grout have widths in the range between about 0.250 in [6 mm] and 0.020 in [0.5 mm]. Continuity of open jointing systems affects penetration: lack of continuity means that more grout holes will be needed than if grout can travel appreciable distances through the systems.

JOINT INCLINATION

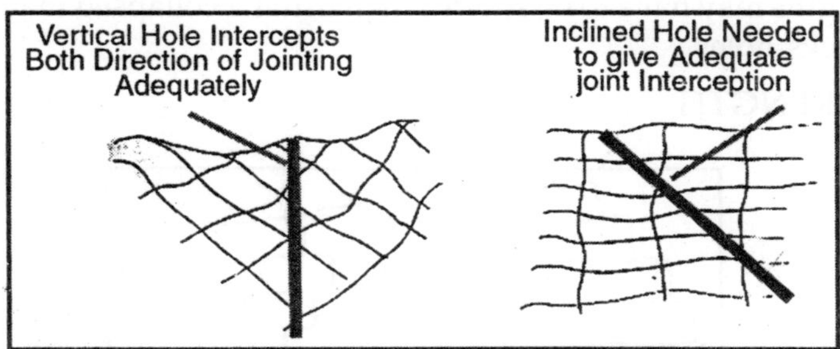

Where dipping is mainly between 20° and about 60°, vertical grout holes may give optimum interception. These are the easiest to drill and are preferable. Steeper jointing usually requires use of inclined holes.

UNIFORMITY OF THE SITE

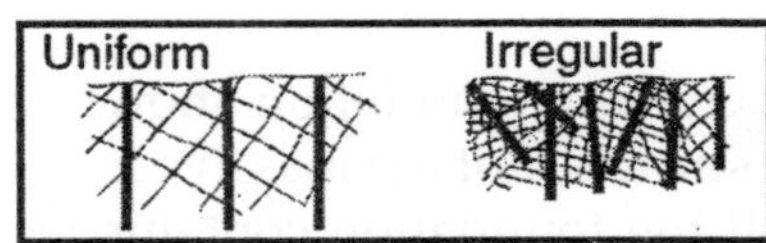

Uniformity of jointing permits a regular layout of grout holes, whereas irregular jointing, dykes, disconformities, and so on may require placement of holes at various inclinations and spacings . Weaknesses may need to be treated especially intensively.

ROCK SOUN2DNESS

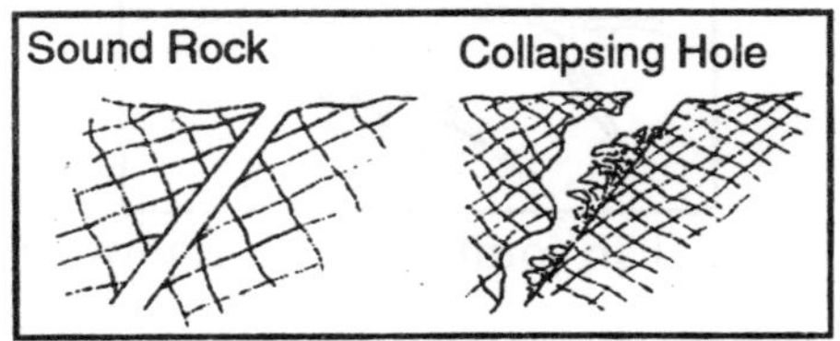

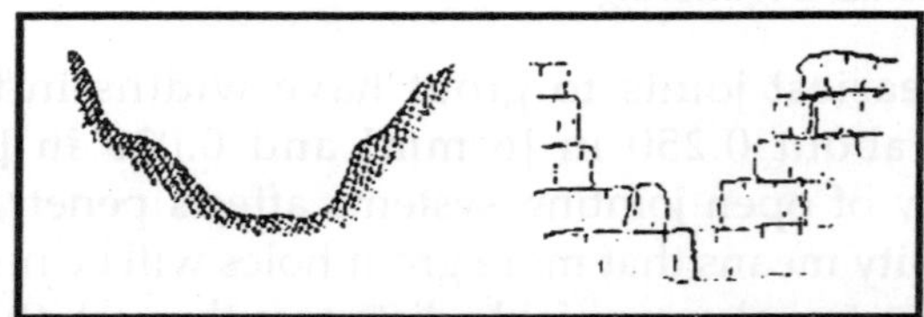

Holes that don't collapse permit easier grouting than those that do. In the latter case, packers cannot be used, and stage lengths may have to be shorter than usual if collapsed material blocks holes.

STRENGTH

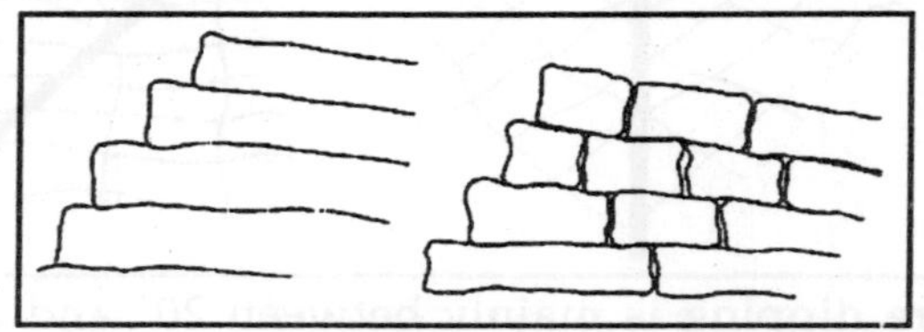

Grouting of strong, massive, well-anchored rock, as sketched at left, is usually easier than when working in weak, broken, loose materials where holes repeatedly collapse, or where blocks move as shown at right.

STRESS IN ROCK

The highly cracked foundation shown at left is unlikely to carry tectonic stresses, but the massive rock sketched at right probably does. If the foundation contains tectonic stress, the grouting will need to take account of this. Early recognition of the condition is necessary.

Piping

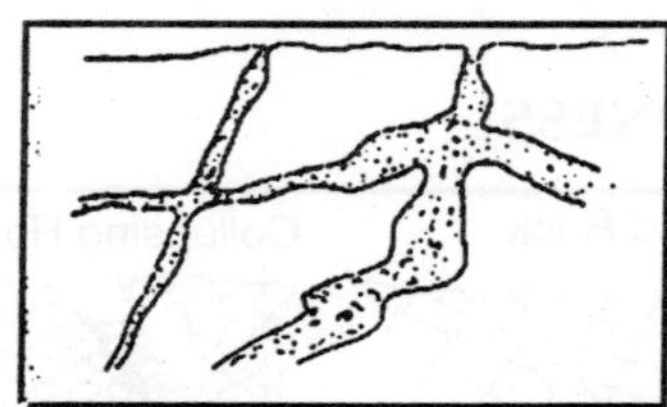

Where material in joints can be removed by seepage,

either by taking the material into solution or by eroding it, the grouting will need to be more intensive than otherwise in order to ensure that seepage through such joints is virtually eliminated.

CHEMICAL ATTACK

The presence of coal or other carbonaceous material or of other deposits that may provide chemically aggressive seepage can warrant provision of a higher standard of grouting than would otherwise be the case.

KARST AND OTHER VOIDS

Large voids such as karst, and also old mines, shafts, and so on, require special provisions when grouted, possibly using fillers in the grout.

STRESSED ROCK

Much of the earth's surface is under stress, usually related to tectonic plate activity. The implications as far as grouting is concerned are that valleys in particular may contain rock under stress that is liable to move under the stimulus of grouting. To get a general idea of how stresses affect a valley, consider the following scenario in geological time.

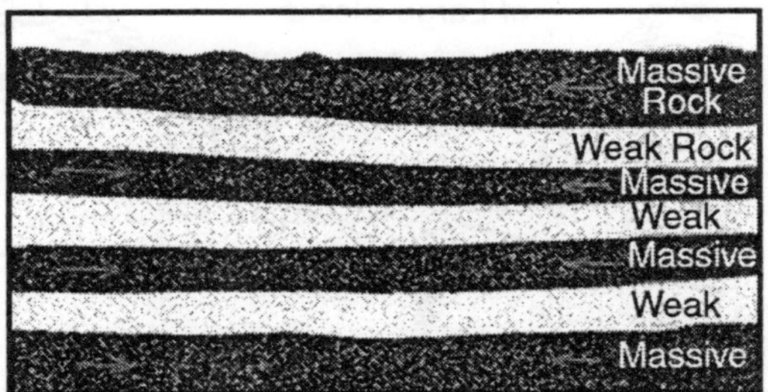

The illustration above shows the process about to be described.

It portrays a foundation comprising alternating beds of massive, strong rock and weak, easily broken material. The massive beds attract any tectonic stresses while the weak beds don't - they shatter and stress relieve themselves to throw off any horizontal forces.

A valley now starts to erode through these beds. After a while the top massive bed is cut and can no longer carry horizontal stresses across the site. Its stresses are then transferred to the next lower massive bed. This transferral takes place well back behind the valley walls, on a regional basis.

However, since this lower massive bed already had stresses in it, it now has more. When erosion eventually proceeds deeper and cuts this highly stressed massive bed its stresses are transferred in turn to the next lower massive bed, which already has its own stresses. The addition of the stresses shed from above may be too much for it and it may fail. This failure may take the form of valley bulging, a local anticline, a fault, or other feature.

Distortion of beds and the presence of highly weathered infill or fault gouge are the usual signs that this has happened. There are easily recognised visual signs if stressed rock is present; they are indicated in the sketch below:

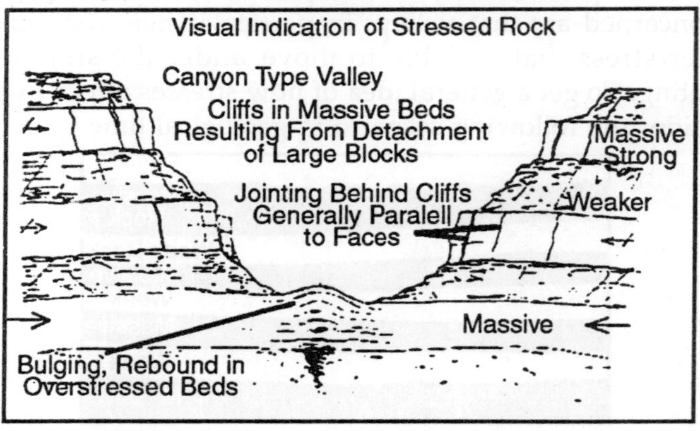

These signs are:

- Valleys tend to be more of the canyon type rather than having gentle slopes.
- The massive beds, indicated by cliff formation, have large blocks bodily detached from the main mass behind. These blocks eventually tumble to expose fresh cliff faces.
- Major open jointing, more or less parallel to cliff

faces, lies behind the faces. This eventually forms the fracture planes of future detached blocks.

- The valley floor has weaknesses as earlier mentioned.
- These visual signs are usually more useful than conducting stress testing and are certainly adequate for grouting purposes.

GROUT CURTAIN DESIGN

Having decided on suitable hole inclinations to suit the various parts of a site, it is then necessary to merge them into an overall design. In doing this some of these carefully considered angles may have to be varied in order to produce a workable design.

Let's start with this simple case. It is a typical dam site valley. For the purposes of the example, jointing is assumed to be approximately horizontal and vertical right across the site. On the left side, grout holes sloping into the abutment at about the inclination shown would give adequate interception of the jointing.

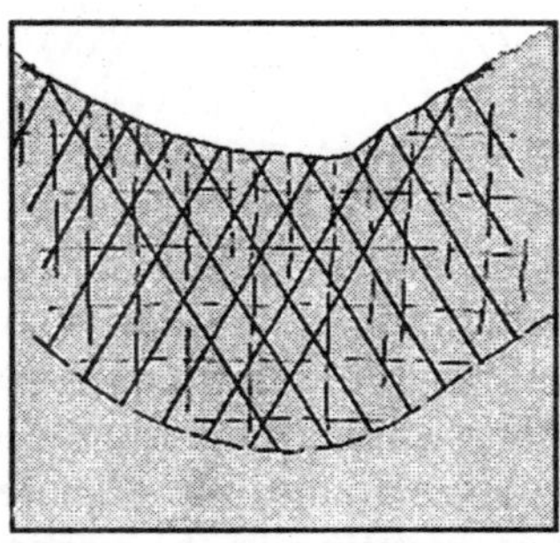

Similarly, the same inclination in the right abutment would be satisfactory. However, the inclinations in the two abutments are in opposite directions to each other; so what should be done where they meet in the floor of the valley?

There are two options:

1. Transition smoothly from one direction to the other as shown in the sketch. Notice how poorly the vertical joints are intercepted. And these are quite liable to carry seepage through the line of the curtain if not properly grouted.

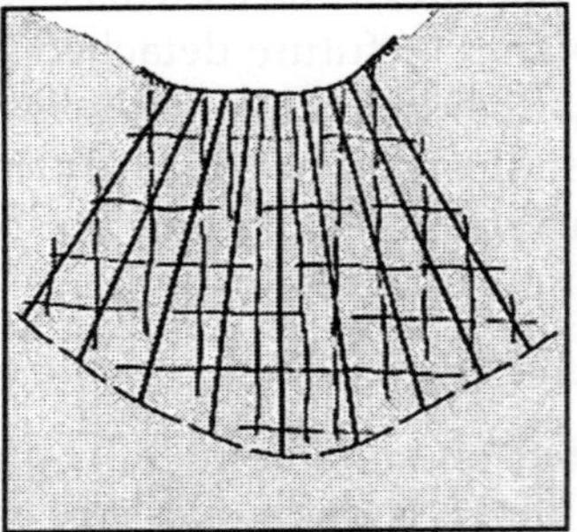

2. Crossover from one direction to the other while maintaining the best inclinations throughout as shown. Vertical cracks are adequately cut in this method, and although it appears to have greater length of holes, it is the recommended method.

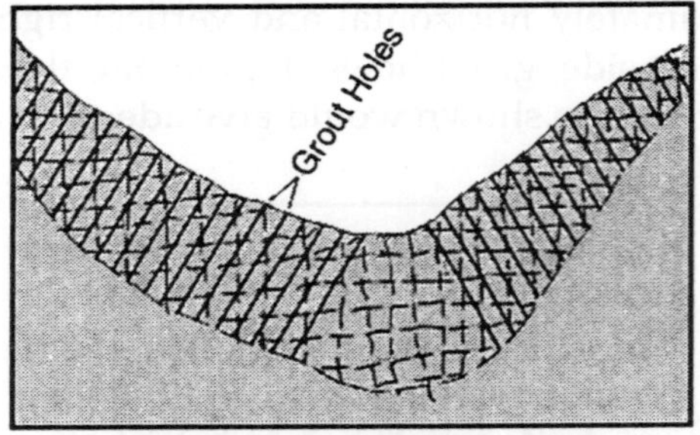

In order to check out both alternatives thoroughly, I grouted several comparable dams using each method. Even though the transition method started with less holes than the crossover method, so many additional holes were needed to bring the standard of grouting in the vertical cracks down to an acceptable figure that the final number of holes exceeded those in the crossover method in all cases, and far more time was lost in fiddling around in confined working areas.

Another important factor favouring use of the crossover method is that it gives doubled-up treatment of geological weaknesses in the valley floor.

Nearly every valley has some weakness in its floor - that is why the valley is there - water has followed the weakness to erode the valley. This may take the form of a fault, shear zone, anticline, graben, or some other feature and presents a specific feature warranting special attention. This is best provided by the crossover layout

Of course, if unlike the previous example, the jointing is amenable to vertical holes, things are simplified in the valley floor. The vertical holes simply continue across - unless weaknesses there specifically require locally inclined holes.

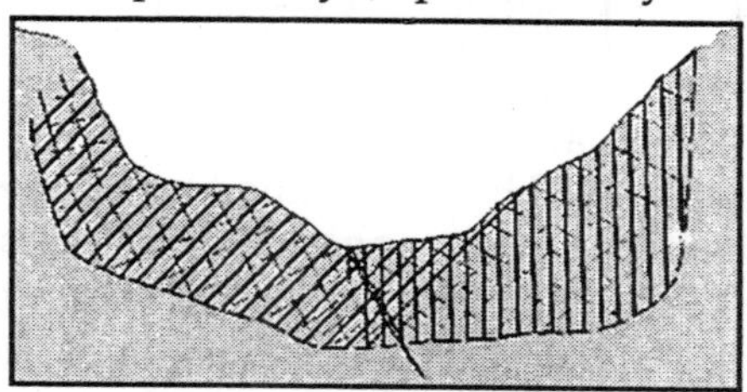

Curtains do not have to be symmetrical. If the jointing in one of the abutments is quite different to the other, hole inclinations should also differ. The sketch shows an example.

The left abutment requires fairly well angled holes, whereas vertical holes will suffice in the right. A fault separates the two and is crossed by holes from both directions in the crossover.

The extreme right-hand hole of the crossover is in an awkward location for drilling; this sometimes happens on real sites but might be avoided by relocating the hole and changing its angle to still provide adequate intersection in the area it deals with, which is down near the base of the curtain. Dips on the left abutment gradually vary, but it is unwise to gradually vary hole inclinations to match (complications and inadvertent errors in hole layout can be introduced if they are). If the curtain happened to be longer than the one shown, a change to another angle for part of it would be warranted.

In this example the curtain has not been deepened in the vicinity of the fault. If water testing had found permeabilities there in excess of the adopted standard for the curtain, then deepening would indeed be needed.

The cases so far have been in valleys where sloping abutments have dictated the direction in those having sloping holes. However, alternative directions are possible if the surface is fairly level.

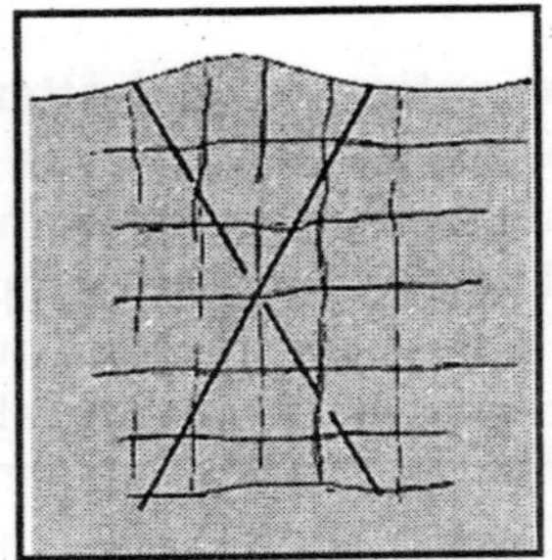

The horizontal and vertical joints shown here can be intercepted equally well by the inclined hole shown in full line or by the hole at the same angle but in opposite direction shown dashed. This concept is used in some multiple-row curtains where one row is at one of the angles and the next is at the other angle. If there are more than two rows, the alternative arrangement continues in each.

EXPERIENCE WITH GROUTING

Grouting is an exercise to decrease the permeability of foundation rocks and soils; as such, its practice is governed by a knowledge of the geology of a site, of the engineering needs of a project, and of the art of testing, selecting and injecting cementitious materials. No college offers such a curriculum; I and any other expert in the field has had to acquire the skills by individual study, research and experience.

In the course of study of geology leading to a master's degree, I learned of the applications of geotechnical engineering, so attained an undergraduate education in civil engineering, preparatory to a three-year job with the hydro-electric department of Cerro de Pasco Corp. in Peru. That mining experience lead me as an aspirant for the Ph.D. in Engineering Science to research the flow of water in fractured rock as a basic element of mine drainage, grouting and foundation engineering. My dissertation was funded by the mining chemicals department of American Cyanamid Co., in order to help market their chemical grouts. My publications on rock fracture properties established a reputation in grouting circles and I was invited to the ASCE Committee on Grouting. At Colourado School of Mines, where I taught groundwater and engineering geology for thirteen years, I authored other papers on grouting and especially fracture hydrology.

Experience working on several dams has included grout treatment and the evaluation of grout effectiveness. These have been important components of consultations on Tablachaca Dam, Peru; Virginia Ranch Dam, Exchequer and Snelling Dams, CA; Kremasta Dam, Greece; and some 40 others of the U.S. Corps of Engineers and Bureau of Reclamation whose grouting history I reviewed for American Cyanamid Co. The grouting records and field conditions of such dams as Kremasta in Greece, Grancarevo in Yugoslavia, Monteynard in France, Hoover and Kerr Dams in U.S., Kariba in Zambia-Rhodesia, Koyna in India and Herdrik Verwoerd in South Africa were reviewed mainly as they revealed foundation properties and conditions.

Similarly, I drew on field and office studies of grout experience at such dams as Oroville, Cherry Creek, Don Pedro and Auburn Dams, California; Grose, Morrow Point, and Horsetooth Dams, Colourado; Grande Coulee, Chief Joseph and Rocky Reach Dams, Washington; and others in writing publications on foundation seepage and grouting. I prepared litigation on the failed Teton Dam, Idaho, which resulted from inadequacies of grout treatment of the cavernous tuff foundations.

In several instances, I have been a consultant on, or a reviewer of tunnel grouting works, including the Rio Blanco Tunnel, Chile; Tooma-Tumut Tunnel, N.S.W., Australia; the Inland Feeder and Cleveland Tunnels, California. I designed seals for the Mount Emmons Mine, Colourado.

Treatment of water inflows to shafts and mines by grouting have been the subject of consulting work and litigation support in recent years, notably that of the flooded K2 Mine, of IMC., Inc. at Esterhazy, Saskatchewan, the flooded shaft at Morton Salt Company's Mines Seleine, Quebec, and the flooding Potocan Mine in New Brunswick.

For City of Ottawa, I investigated causes of failed grouting of the West Rideau Sewer Tunnel and prepared testimony. The great variety of settings represented above suggests that I possess the pertinent geologic background needed to appreciate the problems, procedures and potentialities at any new enterprise.

It is also to be said that I have dealt with essentially all testing and grouting methods and materials, as well as possible substitutes for grouting, such as ground freezing, slurry trenching and dewatering by pumpage. I consider grouting to be but one of the methods of groundwater control, the most general object of my 40 years experience as a consultant to mining and civil engineering firms and agencies.

POZZOLANIC MATERIALS

INTRODUCTION

The use of rendering mortars is an ancient practice that has suffered variations in mortar composition and application related to geographic and temporal factors. The use of pozzolans in mortars is said to have been initiated by the Greeks in 1500 B.C.

Their employment during the Roman period can still be found in ruins widespread throughout the Roman Empire (Rua,1998). Later, during the Byzantine period, they were applied in both Turkey (Oguz Kiliç *et al*, 2004; Çizer *et al*, 2004) and Greece (Moropoulou *et al*, 2004). In volcanic regions, such as the Azores islands in Portugal, these materials

were used in pointing and rendering mortars until recently, when they were replaced by Portland cement. The specificity of lime-pozzolan mortars is their ability to harden under water. This capability enabled their application in Roman baths and tanks, and in maritime structures during the nineteenth century. Although knowledge of pozzolanic mortars is available and a few studies of their characterization have been conducted, their behaviour as renders deserves a further insight. With this intention, two different natural pozzolans were incorporated into lime-based mortars, and testing campaigns were developed to determine their performance when different proportions and curing conditions were used.

THE POZZOLANIC MATERIALS

Natural pozzolans were chosen according to their availability in the Portuguese market; Cape Verde pozzolans had been commercialised and used by the construction industry and pozzolans from the Azores had been used to manufacture pozzolanic cement in the island of S. Miguel. Use of this last material in construction during the 19th Century is vastly documented, especially in the field of coastal defence works.

In terms of visual characterization, Azores pozzolans are of a yellow-brown colour, whilst Cape Verde pozzolans are light grey (Figure).

Although Azores pozzolans have larger particles, both materials were passed through the 0.500 sieve and, for that fraction, national pozzolans present a higher specific surface (6870 cm2/g versus 3250 cm2/g for the other material). XRD analysis of both materials indicates the presence of amorphous material, classified as aluminosilicates.

Fig. Azores Pozzolans

Fig. Cape Verde Pozzolans

Although both materials seem suitable for use as pozzolanic additions, based on this preliminary analysis, tests for pozzolanic reactivity were performed according to the Portuguese legislation (RBLH – Caderno de Encargos para Fornecimento e Recepção de Pozolanas).

These tests, performed at the ages of 7 and 28 days and based on mechanical strength developed by reaction of the pozzolans with lime, classified Cape Verde pozzolans as highly-reactive and Azores pozzolans as low-reactivity.

Neither of these materials complies with the compressive strength requirement of 4,1MPa at 7 days contained in ASTM C 593 Standard Specification for Fly Ash and Other Pozzolans for Use With Lime.

TESTING CAMPAIGN

The results presented in this paper pertain to flexural and compressive strength tests that provide a basic characterization of the mortars' mechanical properties and give a first insight into the mortars' applicability. The testing campaign was developed in three different phases. In the first phase, mortars incorporating both pozzolans were made with a 1:1:4 volumetric ratio (lime:pozzolan:aggregate), which were compared with a lime mortar executed with a commonly-used 1:3 (lime:aggregate) volumetric ratio.

Ratio 1:1:4 was chosen because both carbonation and pozzolanic reaction play important roles in the strength evolution of mortars, and pozzolanic material is composed of reactive and non-reactive phases. A commercial aerial hydrated lime was used and the sand was siliceous river sand.

The use of SEM/EDS techniques provided additional data that may further elucidate the behaviour of lime/pozzolan mortars, by taking into account the possible relation between strength development and the development of CSH in lime/pozzolan mortars. SEM photographs of LCVP2 at an early age (prior to 28 days) and at 3 years of age show a predominance of CSH filaments shortly after mortar production and very few of these structures after 3 years in laboratory conditions (Figure). This fact suggests that the hydraulic components resulting from pozzolanic reactions are unstable, changing quickly with time.

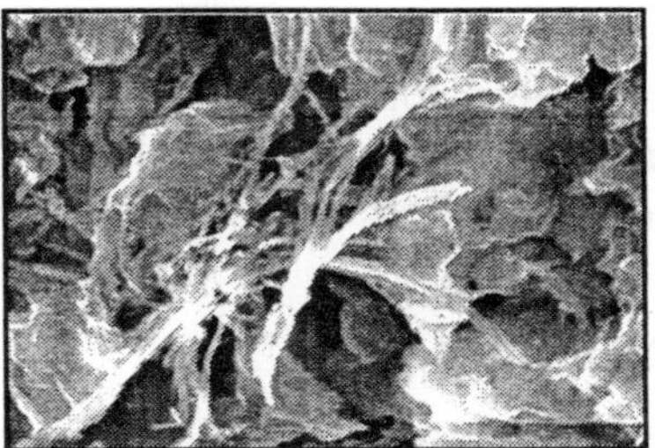

Fig. LCVP2 aged a Few days Showing CSH Filaments

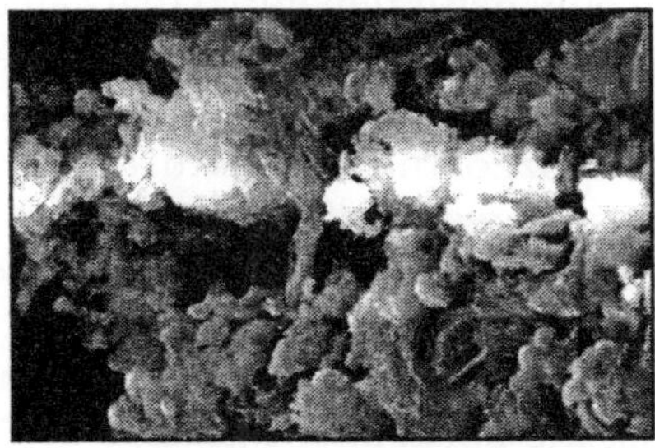

Fig. LCVP2 aged 3 years with Hardly any CSH Filaments

CONCLUSIONS

With these tests, the variation in mechanical strength in mortars with pozzolanic additives became evident. Preparation of mortars with very different characteristics for different ends is a possibility that must be supported by further tests of mechanical and physical characteristics. Decrease of mechanical strength seems to be gradual, but not very accentuated.

This decrease does not jeopardize the use of mortars with adequate strength for application in conservation practice (Veiga, 2003). Furthermore, these mortars seem to attain high durability and capability to endure salt decay, since mortars with pozzolanic additives that were applied in construction in Portugal seawater sustained in good condition. The influence of humidity and initial conditions on the development of strength of mortars with pozzolanic additives seems very significant. Knowledge of the evolution of strength with time and the curing conditions that are possible to obtain in actual work must be taken into account when prescribing the application of these kinds of mortars as renders.

POZZOLANIC MATERIALS AND THEIR ROLE

It's rare to find in nature natural stones with exceptional resistance to every chemical and mechanical corrosion; many silicates have these features and, among these, the natural Pozzolana is the only natural material that can be used as such with lime to form a cement able to harden and resist indefinitely in water.

The natural pozzolana is a natural fine volcanic ash, typical of the volcanic region of Pozzuoli (from here, its name) near Naples, Italy, but present in other volcanic zones of Italy and other Countries.

It was used for the first time by the ancient Romans who started using just the pozzolana extracted at Pozzuoli.

Just the buildings of the ancient Pozzuoli (Puteoli, in Latin) were built using the pozzolana as cement and they resisted for many centuries till today to the action of sea waters that submerged their imposing ruins because of the bradyseism, typical of that area.

The pozzolana is formed by volcanic ashes cemented by the heat of deep magmas and then, after their eruption, exposed to a long action of the atmospheric agents, H_2O and CO_2, so that the original complex silicates were transformed in a fine dust of simple silicates and oxides (SiO_2, Al_2O_3, Fe_2O_3).

When these oxides react with lime, Ca(OH)2, they harden relatively quickly, also under water, forming complex and insoluble calcium silicates and aluminates, more and more resistant with the passing time and not needing to react also with CO_2 to form $CaCO_3$ (limestone) as it happens in the cement, like that used by the Romans before they discovered the pozzolana.

The Romans were able to built exceptional monuments and palaces, like the Pantheon dome in Rome and the foundations of their harbours just with this volcanic cement, the "opus coementicium".

The Roman architect Vitruvium already distinguished 4 types of pozzolana; white, yellow, gray and red.

After the fall of the Roman Empire, the use of this cement was lost and forgotten; buildings like the Pantheon become impossible during the Middle Ages, for the absence of an equally resistant cement to water.

Only Filippo Brunelleschi, the genial Florence architect of the first half of the XV century, started to use again the pozzolana and, since the late Renaissance, this cement has been used frequently for some specific works, reaching again the building quality of the ancient Romans and it's still in use.

Very used is also the ARTIFICIAL pozzolana, prepared heating at about 700C a CLAY mixture formed by SiO2, Al2O3, MgO and CaO.

The natural pozzolana is still extracted today in many volcanic zones but it's replaced, in most of cases, by the HYDRAULIC CEMENT, discovered in England about 1750 and made of LIME and the much more common CLAY, with the same function of pozzolana.

Various mixtures of this new cement were improved during the XIX century, until reaching the formula of the PORTLAND CEMENT (1924), formed by limestone and 40% of clay that hardens with higher speed than pozzolana.

Then, mixtures of the Portland cement with natural or artificial pozzolana started to be available, in the last century, to improve cement resistance to water.

So, pozzolana is again more and more frequent in some cements used for modern restoring works and for building bio-compatible houses.

Chapter 9

Dams, Reservoirs, Tunnels, Bridges and Highway

DAMS AND RESERVOIRS

OVERVIEW OF DAMS AND RESERVOIRS

A dam is any barrier that holds back water; dams are primarily used to save, manage, and/or prevent the flow of excess water into specific regions. In addition, some dams are used to generate hydropower. This article examines man-made dams but dams can also be created by natural causes like mass wasting events or even animals like the beaver.

Another term often used when discussing dams is reservoir. A reservoir is a man-made lake that is primarily used for storing water. They can also be defined as the specific bodies of water formed by the construction of a dam. For example, the Hetch Hetchy Reservoir in California's Yosemite National Park is the body of water created and held back by the O'Shaughnessy Dam.

A Dam is a barrier built to contain water and prevent flooding. The water confined behind such a structure, often forming a lake or reservoir, used for irrigation, as a source of drinking water and for provision of hydroelectric power. Rivers are the cradles of human civilization. The great civilizations have flourished on the banks of various rivers, world over. Eleven rivers and forty-two tributaries bless Goa, which is known as the Tropical Paradise.The idea of building a dam is to control flooding and to and to harness it for hydropower.

Dam also reduces emissions of sulfur dioxide and carbon dioxide; generating electricity equal to about 40 million tons of coal. Small dams do have an advantage in that they take shorter construction time, small initial capital cost, shorter gestation period with early benefits spread out facilities in various regions creating a better development balance within the same constraintof resources etc. Thus they are also required as necessary supplements to big dams.

They are eminently suited for carry over storages and thus impart greater reliability and stability to the system. They provide most effective way of flood regulation and control. Most reliable during drought when small storages do not get filled up and suffer excessive evaporation loss. Thus reliability of large dams is much greater. In drought years the reliability of small dam is very poor when it reduces to less than 10%.

Despite of some drawbacks, the growing population and their need for greater economic development call for more water, in which demand will exceed availability. More storage will be needed to meet the challenges of development and sustainability, but we must take great care for the environment and cultural inheritance and make sure that those people affected by the development of dams will be better off than before the dams were built".

TYPES OF DAMS

Today, there are several different types of dams and the man-made ones are classified by their size and structure. Typically a large dam is classified as being higher than 50-65 feet (15-20 metres) while major dams are those over 492-820 feet (150-250 metres). One of the most common types of major dams is the arch dam. These masonry or concrete dams are ideal for narrow and/or rocky locations because their curved shape easily holds back water via gravity without the need for a lot of construction materials. Arch dams can have one large single arch or they can have multiple small arches separated by concrete buttresses. The Hoover Dam which is on the border of the U.S. states of Arizona and Nevada is an arch dam.

Another type of dam is the buttress dam. These can have multiple arches, but unlike a traditional arch dam, they can be flat as well. Normally buttress dams are made of concrete and feature a series braces called buttresses along the downstream side of the dam to prevent the natural flow of water. The Daniel-Johnson Dam in Quebec, Canada is a multiple arch buttress dam.

In the U.S., the most common type of dam is the embankment dam. These are large dams made out of soil and rock which use their weight to hold back water. To prevent water from moving through them, embankment dams also have a thick waterproof core. The Tarbela Dam in Pakistan is the world's largest embankment dam.

Finally, gravity dams are huge dams that are constructed to hold back water using only their own weight. To do this, they are constructed using extensive amounts of concrete, making them difficult and expensive to build. The Grand Coulee Dam in the U.S. state of Washington is a gravity dam.

TYPES OF RESERVOIRS

Like dams, there are different types of reservoirs as well but they are classified based on their use. The three types are called: a valley dammed reservoir, a bank-side reservoir, and a service reservoir. Bank-side reservoirs are those formed when water is taken from an existing stream or river and stored in a nearby reservoir. Service reservoirs are mainly constructed to store water for later use. They often appear as water towers and other elevated structures.

The first and usually largest type of reservoir is called a valley dammed reservoir. These are reservoirs that are located in narrow valley areas where tremendous amounts of water can be held in by the valley's sides and a dam. The best location for a dam in these types of reservoirs is where it can be built into the valley wall most effectively to form a water tight seal.

To construct a valley dammed reservoir, the river must be diverted, usually through a tunnel, at the start of work. The first step in creating this type of reservoir is the pouring of a strong foundation for the dam, after which construction on the dam itself can begin. These steps can take months to years to complete, depending on the size and complexity of the project. Once finished, the diversion is removed and the river is able to flow freely toward the dam until it gradually fills the reservoir.

DAM CONTROVERSY

In addition to the high cost of construction and river diversion, dams and reservoirs are often controversial projects because of their social and environmental impacts. Dams themselves affect many different ecological components of rivers such as fish migrations, erosion, changes in water temperature and therefore changes in oxygen levels, creating inhospitable environments for many species.

In addition, the creation of a reservoir requires the flooding of large areas of land, at the expense of the natural environment and sometimes villages, towns and small cities. The construction of China's Three Gorges Dam, for example, required the relocation of over one million people and flooded many different archaeological and cultural sites.

MAIN USES OF DAMS AND RESERVOIRS

Despite their controversy, dams and reservoirs serve a number of different functions but one of the largest is to maintain an area's water supply. Many of the world's largest urban areas are supplied with water from rivers that are blocked via dams. San Francisco, California for example, gets the majority of its water supply from the Hetch Hetchy Reservoir via the Hetch Hetchy Aqueduct running from Yosemite to the San Francisco Bay Area.

Another major use of dams is power generation as

hydroelectric power is one of the world's major sources of electricity. Hydropower is generated when the potential energy of the water on the dam drives a water turbine which in then turns a generator and creates electricity. To best make use of the water's power, a common type of hydroelectric dam uses reservoirs with different levels to adjust the amount of energy generated as it is needed. When demand is low for instance, water is held in an upper reservoir and as demand increases, the water is released into a lower reservoir where it spins a turbine.

Some other important uses of dams and reservoirs include a stabilization of water flow and irrigation, flood prevention, water diversion and recreation.

TUNNEL

- To form into a tunnel, or funnel, or to form like a tunnel; as, to tunnel fibrous plants intonests.
- To catch in a tunnel net.
- To make an opening, or a passageway, through or under; as, to tunnel a mountain; to tunnel a river.

ORIGIN

- A vessel with a broad mouth at one end, a pipe or tube at the other, for conveying liquor, fluids, etc, into casks, bottles, or other vessels; a funnel.
- The opening of a chimney for the passage of smoke; a flue; a funnel. *And one great chimney, whose long tunnel thence The smoke forth threw.*
- An artificial passage or archway for conducting canals or railroads under elevated ground for the formation of roads under rivers or canals, and the construction of sewers, drains, and the like.
- (*Science*: chemical) A level passage driven across the measures, or at right angles to veins which it is desired to reach; distinguished from thedrift, or gangway, which is led along the vein when reached by the tunnel.

Tunnel head, the top of a smelting furnace where the materials are put in. Tunnel kiln, a limekiln in which coal is burned, as distinguished from a flame kiln, in which wood or peat is used. Tunnel net, a net with a wide mouth at one] and narrow at the other.

Tunnel pit, Tunnel shaft, a pit or shaft sunk from the top of the ground to the level of a tunnel, for drawing up the earth and stones, for ventilation,lighting, and the like. Tonnelle a semicircular, wagon-headed vault, a tunnel net, an arbor, OF. Also tonnel; dim. Of tonne a tun; so named from its resemblance to a tun in shape

TUNNEL EXCAVATIONS

With the increase in development work of hydroelectric projects in India, the amount of tunnelling work has increased manifold.

Besides to achieve time bound programmes of construction of mega projects in a period of four to five years modern methods of tunnel driving are being considered as only solution to achieve a high rate of progress. In 1980s a progress of 75m per face per month was considered as a high rate of progress whereas nowadays even 150m per face per month is not considered as a good progress. Most of the tunnels now under construction are quite long and normal DBM are failing to achieve a high rate of progress.

The only solution is use of Tunnel Boring Machine (TBMs) to achieve a time bound programme of tunnel excavation in long reaches. To gain confidence level with TBM working in Indian and Himalayan conditions, the systems are yet to be established. Not many tunnels in India have been bored with TBMs, nevertheless it has proved a success in tunnels in India.

Parbati Hydroelectric

The Parbati Hydroelectric Project Stage II, located in Kullu district of Himachal Pradesh, envisages utilisation of water from the Parbati river to obtain a gross head of 862m between the dam site at Pulga and the powerhouse site on the right bank of the Sainj river near Suind village.

A 31.371-km long headrace tunnel and twin pressure shafts with a length of 1,542m convey the water to a powerhouse located above ground. The powerhouse will have an installed capacity of 800 mw. Construction Lot PB.2 includes civil works for 21.233 km of headrace tunnel (HRT) from RD 3500 m to RD 24733 m.

Tunnel from Adit 1 and Adit 3 is being excavated by drill-and-blast method (DBM) whereas 9.050 km of HRT upstream from Adit 2 is to be bored with tunnel boring machine (TBM) from RD 19354 m to RD 10304 m. The excavated diametre of TBM tunnel has been designed as 6.80 m.

Depending on the rock classifications, rock support is envisaged using primarily rock anchors, shotcrete and steel sets. Final lining will be with telescopic folding shutters. Whereas twin pressure shafts as part of Lot PB.3 having length of 1542m are inclined at 30° to the horizontal have been planned to be excavated with TBM.

Pressure shafts shall have an internal diametre of 3.5 m after placement of steel lining. TBM with a nominal excavated diametre of 4.88 m shall be used. According to the technical specifications provided as part of the tender documents, the TBM shall be either an open-type TBM or a single shield TBM. Subsequently contractor has proposed a variant using a double-shield TBM for excavation of the pressure shafts.

Malabar Hill Tunnel, Mumbai

This was the first water supply tunnel driven in India using a hard rock TBM. The tunnel of 3,870m length was driven with 3.5m dia gripper type TBM. The tunnel was reported successfully driven in 450 days with best monthly advance of 376m.

Bandra and Worli sea Outfall Tunnels

Two 3.5m dia tunnels of 3.4 km and 3.7 km length were driven using open face gripper type TBM using precast concrete segmental lining. Tunnel faced unstable ground with a number of rock falls and heavy ingress of water. A peak output of 555m per month was reported in driving Bandra tunnel.

Ghatkopar Sewer Tunnel

This 2,670m long tunnel of 2.5m dia was excavated using fully shielded TBM. Segmental precast concrete lining was used as a support. Many different geological strata including sand, boulders, clay as well as rock were reported to be encountered along with heavy ingress of water. The tunnel was completed in 13 months with best monthly progress reported as 306m and average monthly production of 220m.

Delhi Metro

Construction of Delhi Metro where cut and cover was not feasible is being constructed with two EPB shields and one Dual mode hard rock/shield TBMs. The finished dia of tunnel is 5.70m circular. Where EPB shields are being used the strata comprises of soft soil consisting of sandy-clayey-silt deposits and highly weathered quartzite and mica schist. So far the metro tunnels with EPB shield have been constructed successfully. The length of twin bore tunnels being 4,000m and 916m. Maximum progress with EPB drive reported to be achieved as 332m per month and average being 258m per month.

Dulhasti Hydroelectric Project

The work of Dulhasti tunnel of 8.3m excavated dia tunnel and 6.75km long was started with gripper type hard rock TBM. The rock was predominantly very hard and highly abrasive quartzite. Early problems encountered were high wear of cutters and later on flooding of tunnel twice and burying the whole TBM finally in second flood. This experience in Himalayan geology was not encouraging. The TBM could bore only 2.86km and finally abandoned.

BRIDGES AND HIGHWAYS

Over a period of time, we have acquired skills to manage a large number of Highway and Bridge projects simultaneously with the help of a large pool of in-house experts. We have constructed high value highways and expressways with complex geological condition requiring tunnels, viaducts, flyovers and large span bridges. In India, we are constructing 855 lane kilometres of National Highways and Outer Ring Road Hyderabad besides couple of major bridges and flyovers. Most of our projects are of 8- lane, 6- lane or 4- lane configurations with divided carriageway using flexible and concrete pavements.

We have constructed Cable Stayed Bridges, Arch Bridges, Continuous Box Girder Bridges, Balance Cantilever Bridges, Double Decks Bridges. We have used pre-stressed precast method, Precast segmental methods, post-tensioning precast method, Cast in-situ method besides other methods.

LANDMARK BRIDGES OF INDIA

Project Monitor takes a look at some of the India's most challenging bridge assignments

Jadukata in Meghalaya

Jadukata bridge is situated in the West Khasi Hills district of Meghalaya, some 128 km from Shillong. The most fascinating aspect of the bridge is the 140 m long central span. In fact, it is the longest span cantilever bridge in the country. Contractor: Gammon India Ltd, Mumbai

Dawki Suspension on Meghalaya-Bangladesh border

The Dwaki cable suspension single lane bridge over river Pyan at Dwaki was constructed in 1932. The bridge is Located near Meghalaya-Bangladesh border and is a vital trade link with Bangladesh. Original contractor: Kumardhui Engineering Works

Vidya Sagar Setu in Kolkata

The Vidya Sagar Setu, the second bridge built over river Hooghly, is the longest cable-stayed bridge in India, with a length of 172 m and a main span of 130 m. Contractor: Gammon India Ltd

Narmada Bridge in Gujarat

The 1.4 km long second Narmada bridge on NH-8 in Gujarat was built on BOT basis with advanced techniques such as jack-down method of well-sinking and pre-cast segmental construction. Contractor: Larsen and Toubro Ltd

Yamuna Bridge in Uttar Pradesh

The Yamuna bridge at Kalpi in UP is 767 m long and supports a 7.5 m wide carriageway with a 1.5 m wide footpath on either side. There are eight intermediate spans of 85 m each and two end spans of 43.5m. The superstructure consists of single cell box girders with the deck monolithic with the piers. Contractor: STUP Consultants Ltd

Mahatma Gandhi Setu in Bihar

The Ganga bridge in Patna - now renamed as Mahatma Gandhi Setu - spans over 5.575 km from Hajipur at the north end to Patna at the south end, and is the world's longest river bridge. It was formally opened in March 1982. Contractor: Gammon India Ltd

Brahmaputra Bridge in Assam

The 3,015 m long pre-stressed concrete road bridge over river Brahmaputra was built for North Frontier Railway. This was the second bridge to be built across the highly turbulent river. This bridge provides a vital link between the north-eastern states and the rest of India. Contractor: Hindustan Construction Company Ltd

Teesta Bridge in Sikkim

This is one of the oldest arch bridges built in the north-east in pre-Independent India. The Teesta bridge forms a vital link to Bangladesh, from Sikkim and West Brengal.

Godavari Rail Bridge in Andhra Pradesh

This elegant structure in Andhra Pradesh is the first and only bow string girder arch bridge in the country. The bridge was built for South Central Railway and designed by BBR Bureau, Switzerland. The length of this pre-stressed concrete bridge is 2,745 m. Contractor: Hindustan Construction Company Ltd

New Nizamuddin in New Delhi

The Nizamuddin bridge over river Yamuna on NH-24 interconnects east and west Delhi. Japan extended both grant aid and transfer of technology for the bridge, which included jack-down method for sinking Caisson foundation, pre-casting and launching of girders, gas pressure welded joints and continuity of spans. Contractor: Obayashi Corporation, New Delhi

Chapter 10

Geophysical Exploration

PRINCIPLES

Making, processing, and interpreting measurements of the physical properties of the Earth with the objective of practical application of the findings. Most exploration geophysics is conducted to find commercial accumulations of oil, gas, or other minerals, but geophysical investigations are also employed with engineering objectives, in studies aimed at predicting the nature of the Earth for the foundations of roads, buildings, dams, tunnels, nuclear power plants, and other structures, and in the search for geothermal areas, water resources, archeological ruins, and so on.

Geophysical exploration is also called applied geophysics or geophysical prospecting. The physical properties and effects of subsurface rocks and minerals that can be measured at a distance include density, electrical conductivity, thermal conductivity, magnetism, radioactivity, elasticity, and other properties. Exploration geophysics is often divided into subsidiary fields according to the property being measured, such as magnetic, gravity, seismic, electrical, thermal, or radioactive properties.

MAGNETIC EXPLORATION

Rocks and ores containing magnetic minerals become magnetized by induction in the Earth's magnetic field so that their induced field adds to the Earth's field. Magnetic exploration involves mapping variations in the magnetic field to determine the location, size, and shape of such bodies. The magnetic susceptibility of sedimentary rock is generally orders of magnitude less than that of igneous or metamorphic rock. Consequently, the major magnetic anomalies observed in surveys of sedimentary basins usually result from the underlying basement rocks. Determining the depths of the tops of magnetic bodies is thus a way of estimating the thickness of the sediments.

Except for magnetite and a very few other minerals, mineral ores are only slightly magnetic. However, they are often associated with bodies such as dikes that have magnetic expression so that magnetic anomalies may be associated with minerals empirically. For example, placer gold is often concentrated in stream channels where magnetite is also concentrated.

GRAVITY EXPLORATION

Gravity exploration is based on the law of universal gravitation: the gravitational force between two bodies varies in direct proportion to the product of their masses and in inverse proportion to the square of the distance between them. Because the Earth's density varies from one location to another, the force of gravity varies from place to place. Gravity exploration is concerned with measuring these variations to deduce something about rock masses in the immediate vicinity. Gravity surveys are used more extensively for petroleum exploration than for metallic mineral prospecting. The size of ore bodies is generally small; therefore, the gravity effects are quite small and local despite the fact that there may be large density differences between the ore and its surroundings.

SEISMIC EXPLORATION

Seismic exploration is the predominant geophysical activity. Seismic waves are generated by one of several types of energy sources and detected by arrays of sensitive devices called geophones or hydrophones. The most common measurement made is of the travel times of seismic waves, although attention is being directed increasingly to the amplitude of seismic waves or changes in their frequency content or wave shape.

ELECTRICAL AND ELECTROMAGNETIC EXPLORATION

Variations in the conductivity or capacitance of rocks form the basis of a variety of electrical and electromagnetic exploration methods, which are used primarily in metallic mineral prospecting. Both natural and induced electrical currents are measured. Direct currents and low-frequency alternating currents are measured in ground surveys, and ground and airborne electromagnetic surveys involving the lower radio frequencies are made. See also Geoelectricity.

RADIOACTIVITY EXPLORATION

Natural radiation from the Earth, especially of gamma rays, is measured both in land surveys and airborne surveys. Natural types of radiation are usually absorbed by a few feet of soil cover, so that the observation is often of diffuse equilibrium radiation. The principal radioactive elements are uranium, thorium, and potassium; radioactive exploration has been used primarily in the search for uranium and other ores, such as columbium, which are often associated with them. The Geiger counter and scintillation counter are instruments generally used to detect and measure the radiation.

REMOTE SENSING

Measurements of natural and induced electromagnetic radiation made from high-flying aircraft and earth satellites are referred to collectively as remote sensing. This comprises both the observation of natural radiation in various spectral bands, including both visible and infrared radiation, such as by photography and measurements of the reflectivity of infrared and radar radiation.

WELL LOGGING

A variety of types of geophysical measurements are made in boreholes, including self-potential, electrical conductivity, velocity of seismic waves, natural and induced radioactivity, and temperature variations. Borehole logging is used extensively in petroleum exploration to determine the characteristics of the rocks which the borehole has penetrated, and to a lesser extent in mineral exploration.

Geophysics is a hybrid science (a combination of Geology and Physics) that achieved a distinctive identity only in the mid-twentieth century, which is understandable when it is recalled that neither geology nor physics emerged as distinctive disciplines until the mid-nineteenth century. The antecedents of geophysics reach back to Isaac Newton.

The geophysical exploration of the Americas began with the French expedition of 1735–1745 to the Peruvian Andes, which was led by Charles Marie de La Condamine and included Pierre Bouguer. Paired with a simultaneous venture to northern Scandinavia, the expedition attempted to verify the Newtonian prediction that the earth would be found to bulge at the equator and narrow at the poles, and the expeditions thus inaugurated the study of geodesy (measuring the earth's surface). About sixty years later, between 1799 and 1804, Alexander von Humboldt, accompanied by Aimé Bonpland, explored the American equatorial zone. In a romantic survey of natural history, Humboldt included such broadly geophysical measurements as terrestrial magnetism, which he linked to other physical phenomena—usually meteorological.

These ventures set the style of geophysical exploration in America, establishing precedents for geophysics both as the direct object of an expedition—as in the pursuit of geodesy—and as a component of a larger reconnaissance including Cartography, specimen collection, and geophysical measurements. For nineteenth-century America, the broader Humboldtean model was the more powerful: geophysics was rather an instrumental component of geographic surveys than a conceptual framework for geologic interpretation. After Darwin, this pattern was incorporated into an evolutionary model that supplied the theoretical context of earth science for more than a century.

Concern with applying physical theories and techniques gradually created an informal alliance between three scientific groups. Planetary Astronomy, particularly as practiced by George H. Darwin, Osmond Fisher, and William Thomson (all British), addressed such problems as the formation, age, and structure of the earth. At the same time, "dynamical" geology, in contrast to historical or evolutionary geology, attempted to explain earth processes in terms of mechanics and physical laws.

This group received some inspiration from such meteorologists as James Croll, who fashioned explanations for atmospheric dynamics based on physical processes; what physical chemistry and physical astronomy were to their respective disciplines, dynamical geology was to earth science. Finally, there was mining engineering, which provided technical training in metallurgy, mathematics, and physics at such schools as Columbia University and the University of California. Many of the instruments typical of geophysics, and many of the explorations that used them, stemmed from high-level prospecting, especially for oil.

In the exploration of the American West, geophysics is better understood in terms of certain themes and personalities than as a disciplinary science. Grove Karl Gilbert extended mechanics to problems in geomorphology and structural geology, framing quantitative geologic observations into rational systems of natural laws organized on the principle of dynamic equilibrium. Clarence E. Dutton, elaborating on speculations by John H. Pratt and George B. Airy, conceived the idea of isostasy, or the gravitational equilibrium of the earth's crust, and demonstrated how this pattern of vertical adjustment could become a compressive orogenic force (mountain formation, especially by the folding of the earth's crust). Dutton later made original contributions in volcanology and seismology. Samuel F. Emmons, Clarence King, and George F. Becker applied geochemical and geophysical analysis to the problems of orogeny and igneous ore formation. The latter two men were instrumental in establishing a chemistry laboratory in the U. S. Geological Survey and in applying its experimental results to geophysical phenomena. Becker explicitly attempted mathematical and mechanical models to describe ore genesis and the distribution of stress in the earth's crust.

He was instrumental in the establishment of the Carnegie Institution's Geophysical Laboratory, served as the laboratory's first director, and bequeathed part of his estate to the Smithsonian Institution for geophysical research. Geophysics advanced in the context of a symbiosis (mutually beneficial relationship) of field exploration and laboratory investigation. In the case of the U. S. Geological Survey, Carl Barus staffed the laboratory and Robert S. Woodward furnished field geologists with information on mathematical physics. Woodward later directed the Carnegie Institution.

Following the work of the explorers, other geologists and geodesists—among them, Bailey Willis, Joseph Barrell, and J. F. Hayford—generated quantitative models for earth structure and tectonics. But explanations developed for glacial epochs best epitomized the status of geophysics: attempts to relate glacial movements to astrophysical cycles provided a common ground for geology, geophysics, astronomy, and Meteorology, but the results were rarely integrated successfully. Significantly, the most celebrated attempt at global geophysical explanation remained an unassimilated hybrid. In developing the planetesimal hypothesis in 1904, Thomas C. Chamberlin preserved a naturalistic understanding of earth geology, while F. R. Moulton supplied the mathematical physics. The earth sciences continued to subordinate their data and techniques to a broad evolutionary framework.

By the early twentieth century, geophysics was a conglomerate of pursuits promoted through federal scientific bureaus (Coast and Geodetic Survey, Geological Survey), private or university research institutes (Carnegie Institution's Geophysical Laboratory, the Smithsonian Institution), companies engaged in mineral prospecting, and exceptional individuals. Geophysics, having neither a disciplinary organization nor a unifying theory, remained more an analytic tool than a synthetic science.

This condition persisted until after World War II. Thereafter, with new instruments and techniques developed for mining and military purposes, with additional subjects (especially Oceanography), and with a theoretical topic to organize its research (continental drift), geophysics developed both an identity and a distinctive exploring tradition. This was well exemplified by the International Geophysical Year (IGY), planned for 1957 and 1958 but extended to 1959. In counterpoint to the space programme, geophysicists proposed to drill into the interior of the earth. Although aborted in 1963, Project Mohole was superseded by other oceanic drill projects, especially the Joint Oceanographic Institute's Deep Earth Sampling Programme (JOIDES) begun in 1964. The International Upper Mantle Project (1968–1972) formed a bridge between IGY and research under the multinational Geodynamics Project (1974–1979), which proposed to discover the force behind crustal movements.

In the late 1990s, geophysical exploration led to the Ocean Drilling Programme. The programme functioned from a ship called the JOIDES Resolution, named for the Joint Oceanographic Institutions for Deep Earth Sampling. Built to drill into the seabed for oil, the ship's high-technology laboratory was used by an international crew of scientists to conduct geophysical research. The programme found evidence related to the impact of a meteorite at the end of the Cretaceous Era, evidence of ocean temperature changes in the Ice Ages, and documented changes in the earth's magnetic poles.

Geophysics is a revolution in physics, and its integrative concept, the theory of plate tectonics (formerly continental drift), rivals relativity and quantum mechanics in significance. Involving geophysical research in practically all fields of earth science, and paired with satellite surveys, plate tectonics constitutes a new inventory of natural resources and a scientific synthesis of the globe.

Geophysical exploration has, moreover, preserved its archetypal (original) forms, being international and corporate in composition, global in scale, and quantitative in data and being founded on the theoretical assumption of a steady state.

It blends the styles of La Condamine and Humboldt, combining specific geophysical pursuits against a cosmic landscape. Yet the transformation is remarkable—the difference between Humboldt's surveying of sublime panoramas from the summit of the Andes, and the Earth Technology Resource Satellite (ETRS) radioing instrumental data to terrestrial computers.

GEOPHYSICAL METHODS

- This guide covers the selection of surface geophysical methods, as commonly applied to geologic, geotechnical, hydrologic, and environmental investigations (hereafter referred to as site characterization), as well as forensic and archaeological applications. This guide does not describe the specific procedures for conducting geophysical surveys. Individual guides are being developed for each surface geophysical method.
- Surface geophysical methods yield direct and indirect measurements of the physical properties of soil and rock and pore fluids, as well as buried objects.
- The geophysical methods presented in this guide are regularly used and have been proven effective for hydrologic, geologic, geotechnical, and hazardous waste site assessments.
- This guide provides an overview of applications for which surface geophysical methods are appropriate. It does not address the details of the theory underlying specific methods, field procedures, or interpretation of the data. Numerous references are included for that purpose and are considered an essential part of this guide. It is recommended that the user of this guide be familiar with the references cited and with Guides D 420, D 5730, D 5753, D 5777, and D 6285, as well as Practices D 5088, D 5608, D 6235, and Test Method G 57.
- To obtain detailed information on specific

geophysical methods, ASTM standards, other publications, and references cited in this guide, should be consulted.

- The success of a geophysical survey is dependent upon many factors. One of the most important factors is the competence of the person(s) responsible for planning, carrying out the survey, and interpreting the data. An understanding of the method's theory, field procedures, and interpretation along with an understanding of the site geology, is necessary to successfully complete a survey. Personnel not having specialized training or experience should be cautious about using geophysical methods and should solicit assistance from qualified practitioners.
- The values stated in SI units are to be regarded as the guide. The values given in parentheses are for information only.
- This guide offers an organized collection of information or a series of options and does not recommend a specific course of action. This document cannot replace education or experience and should be used in conjunction with professional judgment. Not all aspects of this guide may be applicable in all circumstances. This ASTM standard is not intended to represent or replace the standard of care by which the adequacy of a given professional service must be judged, nor should this document be applied without consideration of a project's many unique aspects. The word "Standard" in the title of this document means only that the document has been approved through the ASTM consensus process.

This standard does not purport to address all of the safety concerns, if any, associated with its use. It is the responsibility of the user of this standard to establish appropriate safety and health practices and determine the applicability of regulatory limitations prior to use.

Ever since the early textbooks were written on the applications of geophysics, seismic exploration methods have been divided into the reflection and refraction techniques. In each technique, a number, from at least 12 to as many as hundreds, of readily portable seismometres (geophones) are laid out along a line (or lines in the case of 3-dimensional seismic reflection studies), and a number of seismic sources are energized in sequence. The methods used for recording the signals from these geophones can vary greatly, as can the character of the seismic source. Because the early practice was to use small explosive charges, the source is consequently thought of as a shot that generates a distinct package of seismic body wave energy: a wavelet which travels by various ray paths through the Earth and eventually returns to the surface to be recorded by the geophones. The travel times and amplitudes of these wavelets can be interpreted to reveal the subsurface structure.

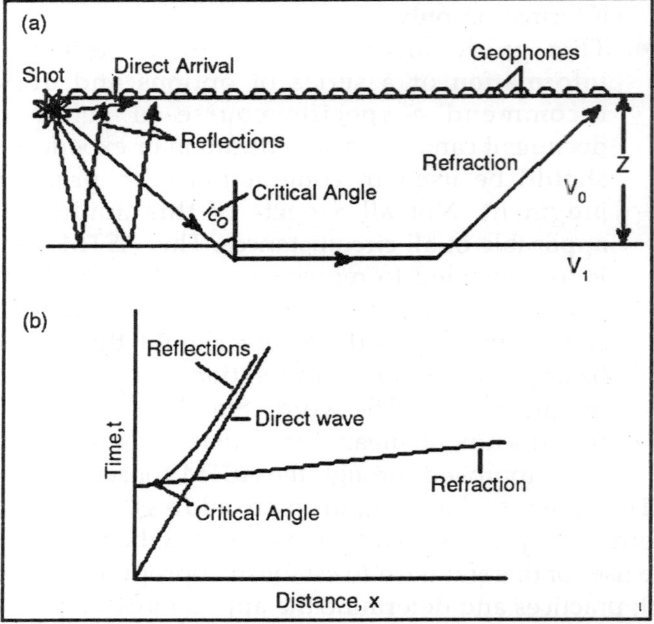

The reflection technique is dominant because it can produce a very detailed image of the subsurface structure by recording nearly vertical travelling waves as they are reflected from discontinuities in the Earth (Fig. 1a). The refraction technique, on the other hand, targets waves which have a large horizontal component to their travel path. Earth structure is then deduced from computer modelling of the observed data by employing techniques such as ray tracing.

This division into two techniques has historically made sense because of the different approaches that are used to interpret the data and because of the limitations imposed by instrumentation. However, in either case, body waves are employed and both reflected and refracted waves are recorded and used. Either technique is very flexible and can be employed in near-surface studies for engineering and environmental applications or in studies of deep Earth structure. If the number of geophones that can be deployed is limited to, say, 48, the choice is either to place the instruments close together, as is done in the reflection technique, or spread them apart, as is done in the refraction technique. Thus, the perception is that the reflection technique produces a useful image of the Earth but has limited resolution for determining velocity, while the refraction technique measures velocity well but can produce only simplistic models of the Earth. However, with advances in instrumentation that allow hundreds or even thousands of geophones to be deployed, and with the development of new theoretical approaches, this division is becoming blurred. Because waves travelling by a wide variety of travel paths can all be densely recorded, the advantages of both ways of treating the problem can begin to be exploited simultaneously.

Index

H

I

L

M

N

O

R

S